TOPICS IN MICROMETEOROLOGY

TOPICS IN MICROMETEOROLOGY. A FESTSCHRIFT FOR ARCH DYER.

Edited by

BRUCE B. HICKS

NOAA/ARL, Turbulence and Diffusion Division,
Oak Ridge, Tennessee, U.S.A.

Reprinted from Boundary-Layer Meteorology
Vol. 42, Nos. 1–2 (1988)

D. Reidel Publishing Company / Dordrecht / Boston

Library of Congress Cataloging in Publication Data

CIP

CIP-data appear on seperate card.
ISBN 90–277–2694–9

Published by D. Reidel Publishing Company,
P.O. Box 17, 3300 AA Dordrecht, Holland.

Sold and distributed in the U.S.A. and Canada
by Kluwer Academic Publishers,
101 Philip Drive, Norwell, MA 02061, U.S.A.

In all other countries, sold and distributed
by Kluwer Academic Publishers Group,
P.O. Box 322, 3300 AH Dordrecht, Holland.

Printed in Belgium

ARCH

Some Introductory Notes to an Issue of Boundary-Layer Meteorology Dedicated to Arthur James Dyer

B. B. HICKS

NOAA Atmospheric Turbulence and Diffusion Division, P.O. Box 2456, Oak Ridge, TN 37831, U.S.A.

This issue is a Festschrift to Dr A. J. ('Arch') Dyer, the distinguished Australian micrometeorologist. In these introductory notes, the intent is to summarize Arch's contributions in micrometeorology, and especially in the development of flux-gradient relationships, not in depth but in sufficient detail to give the unfamiliar reader an idea of the thoroughness of his work. Arch's own review of the history of eddy flux measurement makes interesting reading; in it, his own role is characteristically downplayed (Dyer, 1968).

Arthur James Dyer was born in 1925. He eventually became a student at the University of Melbourne, after leaving an intimidating academic record behind him at Melbourne's University High School. In 1946, Arch was awarded his B.Sc. degree in Physics, with First Class Honours. He went on to receive his M.Sc. degree in 1948, and a Ph.D. in Physics in 1953.

Arch's thesis was on the detection and measurement of cosmic rays. There was no meteorology in his curriculum, but immediately upon joining the Commonwealth Scientific and Industrial Research Organisation's Meteorological Physics Section in 1954, he was rudely introduced to meteorology and especially to the technical limitations of meteorological sensing. At that time, the Section (later to become the Division of Meteorological Physics, then the Division of Atmospheric Physics, and now the Division of Atmospheric Research) was geared towards answering some specific questions about the dependence of surface fluxes on atmospheric and surface conditions. The team Arch joined was powerful, headed by C. H. B. (Bill) Priestley and W. C. (Bill) Swinbank, and with R. J. (Reg) Taylor, E. L. (Len) Deacon, I. C. (Ian) McIlroy, and E. K. (Eric) Webb forming the core of the experimental effort. Arch took over the responsibility for the electronics shop, which was then heavily involved in developments related to eddy correlation.

It is worth remembering the times. Much earlier, Scrase (1930) had demonstrated that eddy correlation could be made to work, but had shown (with great emphasis) the difficulty of the job. In Australia, Swinbank, Priestley, Taylor, and Webb had explored the approach further, and had started intensive work on automating the analysis. The first computerized system employed a mechanical integration device, a ball-and-disc integrator in which one signal controls the speed of rotation of a disc and another controls the radius at which a tangential wheel contacts the disc (Taylor and Webb, 1955). The number of turns of the contacting wheel is directly related to the integral of

Boundary-Layer Meteorology **42** (1988) 1–8.

the product of the two signals. Using three of these systems, it is possible to solve the eddy correlation problem: one system is used to integrate the product of vertical velocity and temperature, say, and two other systems are used to integrate the individual signals (see Swinbank, 1951, 1955). The problem that most severely constrained this delightfully direct approach was the same as that which has kept cropping up ever since – the need for rapid response of the data analysis system as well as of the sensors.

Arch and his team contributed a major improvement: the first fast-response electronic product integrator (Dyer, 1958). This was a modified Watt-hour meter, in which a disc is rotated in response to induced eddy currents in the presence of external magnetic and electrical fields. The number of rotations of the disc is then directly related to the product of the magnetic and electrical field strengths, both of which are modulated by input signals. This innovation formed the foundation for work on eddy correlation until the development of electronic integrators in the mid-1960's. The 'Evapotron' was a field-portable instrument that used three of these integrators in the same way as the ball-and-disc devices that had been used previously (Taylor, 1956; Dyer and Taylor, 1959; Dyer, 1961; Dyer and Maher, 1965).

The team of Taylor and Dyer contributed in the development of sensors, as well. This was well before the era of sonic anemometry, although sonic techniques were among the many potential sensing systems that were investigated. At that time, electronics were not yet well enough developed to make sonic anemometers feasible (likewise for pressure-sensing devices – the other approach that was extensively explored without great profit), and hot-wire technology was the best available. Taylor and Dyer invested enormous energy into the improvement of hot-wire technology; a highly refined heated-wake anemometer was the result (Dyer, 1960). This provided one-second response times, with excellent cosine response.

A major limitation was never overcome with complete satisfaction: as with all hot wire anemometers, the devices did not work well in light winds and, therefore, could not be configured to measure u and w directly; instead, they were arranged to measure $(u + w)$ and $(u - w)$, from which signals proportional to u and w were then extracted. (As usual, u is the longitudinal wind component, w the vertical.) Electronic noise arising in either of the original inputs would then appear in both of the derived velocity components, obviously correlated, and clearly confounding any attempt to derive estimates of momentum covariances. The Evapotron was never fully successful as a means for measuring u_*, but it worked well for sensible and latent heat fluxes.

With a working data-reduction and analysis system and a suitable anemometer for measuring fluctuations in the vertical velocity component, it was then necessary to measure temperature and water vapor. Arch and his team developed a method for standardizing fine-wire thermometers, so that these sensitive elements could be replaced quickly in the field (Dyer and Maher, 1965). The selection of nickel wire was unusual in micrometeorological research, but half-hardened nickel worked well, was easy to handle, and offered a large temperature coefficient. It had an additional advantage of wetability; an element constructed of cotton wound with the nickel wire as an integral element of the thread formed an excellent fast-response wet bulb, which could then be fine-tuned to precisely the same temperature coefficients as the dry bulb. The Evapotron used such wet and dry bulb systems. When operated with a wet bulb, the Evapotron measured $H + LE$, and with a dry bulb it measured H. Using both, it was possible to quantify both H and LE using a field-portable instrument.

The complexity of the sensors and the computing routines used to analyze their outputs created a healthy respect for the problem of error propagation. Fundamental tests were devised to generate confidence in the measurements that were made in the field. The matter of sensitivity to sensor orientation was addressed directly: two sensors were operated side by side, and one was rotated relative to the other. Sensors were operated at different heights to find out what errors would arise if the instruments were operated too close to the surface. Studies were also conducted on the role of fetch, to quantify the error margin associated with operations over fields of limited extent (Dyer and Crawford, 1965). This last study was accompanied by an extensive theoretical analysis of the leading-edge question (Dyer, 1963).

In all such studies, a key consideration was the need to satisfy the constraints of energy balance at the surface. On more than one occasion, it was found that the eddy fluxes were not agreeing well with measurements of net radiation and ground heat

Fig. 1. Arch Dyer, and some memories.

transfer. The entire experimental team packed up, and went home. The lessons learned were strong, and remain a part of good micrometeorological field practice. Any researcher who steps into the field with sensors and recording apparatus which he expects to work well, and does not take steps to reduce data on the spot for real-time inspection and quality control, is setting himself up for some shocks when the analysis is finally completed.

A parallel concern centered around the need to assure that the measured eddy fluxes were correct. This worry generated a series of comparisons against different techniques, such as weighed lysimeters for testing measurements of evaporation rate, and Bowen ratio methods for testing the heat fluxes. The critical measurements were made at a field site near the Aspendale location of the Division of Meteorological Physics (which it was, by this time), at Edithvale. This was a field site suitable for testing and improving equipment, but significantly affected by mobile roughness elements (bovine) over which little control was possible. More satisfactory field sites were sought for studies of flux-gradient relationships, the elucidation of which was the final goal of these developments. There followed the sequence of field studies that blazed the trail for micrometeorological experiments to follow: Deniliquin (western New South Wales) in 1961; Kerang (northern Victoria) three times in 1962; back to Kerang in 1963, and a trial study at Hay (north of Deniliquin, in NSW); Hay once again in 1964, and several times in 1965; and finally to Gurley in northern New South Wales.

Fig. 2. The 1965 site at Hay, N.S.W., showing a few of the instrument towers that were used.

In these experiments, site quality was a basic requirement, to be assessed by studies of the wind profile in neutral stability, and by measurement of horizontal gradients of wind, temperature, and humidity. (A photograph of the site used at Hay, in 1965, is given as Figure 2. Instruments were distributed over a large expanse of very flat grazing land, typically supporting one or two sheep per hectare.) Stationarity was also a fundamental requirement; measurements were not made when conditions were changing with time, such as near dawn and dusk. In all these early studies, it was assumed that the value of the von Kármán constant was 0.4, until measurements of momentum flux improved sufficiently to cause this estimate to be refined to 0.41 (but see the paper by J. A. Businger in this issue). The series of experiments is summarized by Swinbank and Dyer (1968).

In a typical field experiment, three sets of eddy flux instruments would be set up near each other, one each for sensible heat, latent heat of evaporation, and momentum flux. The three-headed monster was affectionately known as 'Cerberus', after the mythical creature guarding the entrance to Hades. The three operators were the same through most of these experiments: Arch Dyer, Guntis Grauze, and Bruce Hicks, the present author. In later years, Peter Hyson joined this team.

The data would be collected and checked as they came off the machines, reduced, scrutinized, and tabulated. Each night, there would be a wake as the day's products were evaluated. Points would be carefully drawn on a plot of dimensionless gradients against stability. It was in some motel room at some stage through these expeditions that a first cautious curve was drawn through a set of points describing the average relationship between eddy fluxes and gradients. The first question was whether $Kh = Km$; the answer was clearly negative. The second question was whether Kw was more like Kh or Km; the data pointed to Kh. The last question concerned the best formulation of each of these eddy diffusivities in terms of the friction velocity, height, and stability. Dyer (1968) and Dyer and Hicks (1970) summarized the results.

During this sequence of field studies the advantages of signal filtering became a center of attention. The subject coincided with the development of the helicoid propeller anemometer as a w sensor. The 'Fluxatron' was developed and was tested in 1965 (Dyer *et al.*, 1967). In this device, the average vertical velocity signal was driven to zero with a time constant that was variously set between 20 and 200 s (this was found to be not critical for unstable conditions, within the accuracy limitations of the available data). The temperature signal was not treated so well; a crude average was removed using a lagged temperature element in a second arm of the temperature bridge circuit. The desired product was obtained by driving the temperature bridge with a signal proportional to w' (the departure from the mean of the vertical velocity). The Fluxatron measured sensible heat only. Shortly thereafter, the development of precise analog multipliers enabled the Fluxatron approach to be generalized, so that running means could be removed from any signal, and permitting computation of any of the many covariances of interest. This generalized covariance computer was the backbone of field studies until modern digital computing techniques took over, in the mid 1970's. The sensing head of the covariance system used in field studies during the late 1960's and early 1970's is seen in the foreground of Figure 3. In the background is a wind profile

Fig. 3. A 'Fluxatron' sensing head, as used in field experiments of the late 1960's and early 1970's. In the background is a wind profile mast.

mast as was used in all of this series of micrometeorological field experiments.

Australian patent number 407427 is held by Arch Dyer. It refers to 'Flux measuring apparatus', and was issued on November 7, 1970.

The developments of new instruments were paralleled by intercomparison studies. The first such study with multinational overtones was conducted at Hay in 1965 (Businger *et al.*, 1967). Arch later played a leading role in the Soviet turbulence sensor comparison study at Tsimlyansk, near the Black Sea, in 1970. (He is fondly remembered by us all for a moving fairwell address he delivered at the final dinner. The speech was delivered flawlessly, in Russian.) As his grand finale to the international experiment scene, Arch organized and hosted the 1976 International Turbulence Comparison Experiment, back in the familiar territory of western NSW, at Conargo, near Deniliquin.

Along the way, Arch detoured into air-sea interaction, and worked extensively on the refinement of high-frequency turbulence intensity methods for determining eddy fluxes at sea. In addition, he was a leading figure among the small cadre of researchers studying radioactive fallout during the 1950's and 1960's. This last item constituted more than just an additional interest. It grew into a second major scientific effort of Arch's, changing with time into a strong interest in stratospheric diffusion and finally evolving into considerations of possible climate change. A review of Arch's publications reveals about 50 papers on micrometeorology, and about 20 on his more global interests. His publications earned him a D.Sc. degree, awarded in 1968 by the University of Melbourne. His papers on flux-gradient relations earned him the Royal Meteorological Society's Buchan prize (also in 1968), shared with W. C. Swinbank. In 1971, he was the recipient of Melbourne University's David Syme research prize for Physics.

In the later years of his meteorological career, Arch was Assistant Chief of the CSIRO Division of Atmospheric Research. He was a Fellow of the Royal Meteorological Society (and was the prime mover behind the setting up of an Australian Branch, of which he became Chairman in 1974), and was a foundation member and Fellow of the Australian Institute of Physics. He was a hard-working member of the Editorial Board of *Boundary-Layer Meteorology*.

The papers collected in this volume are selected for their relevance to Arch's micrometeorological interests. The paper by Businger, for example, addresses directly the matter of the difference between the Australian flux-profile relationships and the North American results based on observations at Kansas. The question of sensor interferences with the flows that are meant to be measured by it is addressed in the paper by Wyngaard; Arch spent a considerable time reflecting on this matter, both experimentally and theoretically (see Dyer, 1981). The results of a recent field experiment that follows the lines of the Australian sequence are presented in the paper by Hogstrum. Other papers show typical applications of the flux gradient relations. Finally, three papers are included to show the way in which the initial work by Dyer and his colleagues on flux measurement techniques has evolved recently, as related to aircraft and towers.

As a result of a bout with ill-health, Arch has been forced to withdraw from the micrometeorological community. He is now concentrating his energies on his artistic interests. Those of us who remember him as a leader of the international community of experimental meteorologists can only regret that he is no longer an active member of our fraternity, but in the same breath we must marvel that any person should have the multidisciplinary talent to enable him to leave one profession and pick up in another, without pause.

Arch's musical talent has always been obvious to those who have been fortunate enough to know him well. Jack Fina's 'Bumble Boogie' version of Rimsky-Korsakov's 'Flight of the Bumble Bee' has a special significance. At some time during the many field experiments in which he participated, Arch Dyer was usually pursuaded to belt out his rendition of this old favorite on whatever poorly tuned piano was then available. At clubs in outback Australian towns such as Kerang in northern Victoria, and Deniliquin and Hay in the western semi-desert of New South Wales, bars would go silent as Arch would monopolize everybody's attention, friends and strangers alike, with his rendition at the keyboard.

A personal note is surely appropriate. The editor of this issue was a young, inexperienced university graduate when Arch took him and guided him into his present career. We all thank you, Arch, but I owe you more than most.

References

Businger, J. A., Miyake, M., Dyer, A. J., and Bradley, E. F.: 1967, 'On the Direct Determination of the Turbulent Heat Flux Near the Ground', *J. Appl. Meteorol.* **6**, 1025–1032.

Dyer, A. J.: 1958, 'An Improved Electromagnetic Integrator', *J. Sci. Instr.* **35**, 240–242.

Dyer, A. J.: 1960, 'Heat Transport Anemometer of High Stability', *J. Sci. Instr.* **37**, 166–169.

Dyer, A. J.: 1961, 'Measurements of Evaporation and Heat Transfer in the Lower Atmosphere by an Automatic Eddy-Correlation Technique', *Quart. J. Roy. Meteorol. Soc.* **87**, 401–412.

Dyer, A. J.: 1963, 'The Adjustment of Profiles and Eddy Fluxes', *Quart. J. Roy. Meteorol. Soc.* **89**, 276–280.

Dyer, A. J.: 1968, 'A Historical Development of Eddy-Flux Measurements', in *Memorial Volume to Professor Frederck A. Brooks: 1895–1967,* 16 pp.

Dyer, A. J.: 1981, 'Flow Distribution by Supporting Structures', *Boundary-Layer Meteorol.* **20**, 243–251.

Dyer, A. J. and Crawford, T. V.: 1965, 'Observations of the Modification of the Microclimate at a Leading Edge', *Quart. J. Roy. Meteorol. Soc.* **91**, 345–348.

Dyer, A. J. and Hicks, B. B.: 1970, 'Flux-Gradient Relationships in the Constant Flux Layer', *Quart. J. Roy. Meteorol. Soc.* **96**, 715–721.

Dyer, A. J., Hicks, B. B., and King, K. M.: 1967, 'The Fluxatron – A Revised Approach to the Measurement of Eddy Fluxes in the Lower Atmosphere', *J. Appl. Meteorol.* **6**, 408–413.

Dyer, A. J. and Maher, F. J.: 1965, *Automatic Eddy Flux Measurement with the Evaporation,* CSIRO Div. Meteorol. Physics, Tech. Paper No. 15, 31 pp. Dyer, A. J. and Taylor, R. J.: 1959, 'The Direct Measurement of Natural Evaporation', *Meteorology* **13**, 143–155.

Scrase, F. J.: 1930, *Some Characteristics of Eddy Motion in the Atmosphere*, U.K. Meteorological Office Geophysical Memoirs, No. 52, 16 pp.

Swinbank, W. C.: 1951, 'The Measurement of Vertical Transfer of Heat and Water Vapour and Momentum in the Lower Atmosphere with Some Results', *J. Meteorol.* **8**, 135–145.

Swinbank, W. C.: 1955, *An Experimental Study of Eddy Transports in the Lower Atmosphere*, Technical Paper No. 20, CSIRO Division of Meteorological Physics.

Swinbank, W. C. and Dyer, A. J.: 1968, *Micrometeorological Expeditions; 1962–1964,* CSIRO Div. Meteorol. Phys. Tech. Paper No. 17, 48 pp.

Taylor, R. J.: 1956, 'Some Measurements of Heat Flux at Large Negative Richardson Number', *Quart. J. Roy. Meteorol. Soc.* **82**, 89–91.

Taylor, R. J. and Webb, E. K.: 1955, *A Mechanical Computer for Micrometeorological Research*, Technical Paper No. 6, CSIRO Division of Meteorological Physics.

THE STREAMWISE KOLMOGOROFF CONSTANT

E. L. DEACON

4 Haldane Street, Beaumaris, Victoria, 3193, Australia

(Received 14 July, 1986)

Abstract. Observations over grassland of the turbulent kinetic energy in a band of frequencies in the inertial subrange of the spectrum of the streamwise wind component are related to the stress indicated by the wind profiles. The object is to determine the effective Kolmogoroff constant, α_{UB}, which accords with the assumption of balance between turbulent energy production and dissipation. The mean from 60 half-hour runs made under stability conditions ranging from neutral to moderate instability is $\alpha_{UB} = 0.62$.

This result is compared with those from other studies; four over grassland and six over the sea or a lake. There is no significant difference between the means of the land and sea values, but the latter are more scattered, partly because of difficulties in securing suitable exposure of the instruments for the stress measurements. The mean value from all ten sets of observations is $\alpha_{UB} = 0.59 \pm 0.025$. So the dissipation method should be capable of giving the drag coefficient of the sea in strong winds with an uncertainty of no more than about 10%.

1. Introduction

During a micrometeorological expedition of the CSIRO Division of Atmospheric Physics to an extensive level treeless site near Hay, New South Wales, in November 1968, trials were made of equipment designed to test the value of the dissipation (or spectral density) technique for the determination of momentum flux. Sixty successful runs were made along with simultaneous observations of wind and temperature profiles and heat fluxes. Unfortunately an essential part of the data was mislaid and only recently rediscovered. As there are some novel features of the apparatus in its Mark III form and the experimental scatter of the results is satisfactorily small, it is thought worthwhile, even now, to give the following report. Furthermore, subsequent published studies permit evaluation of the shear stress from the wind profiles to be made with greater confidence than was possible in 1968. Such estimation was necessary as eddy covariance measurements of momentum flux were not available on this expedition.

The relationship under study is that for the streamwise component of the turbulent motion in the inertial subrange of the spectrum. On the assumption that the rate of dissipation of turbulent kinetic energy is equal to the rate of production by shear and buoyancy, the expression is:

$$\alpha_{UB} = \frac{(2\pi k)^{2/3} f^{5/3} F_{uu}(f)}{u_*^2(\phi_m - z/L)^{2/3}} \tag{1}$$

in which α_{UB} = Kolmogoroff constant for the streamwise velocity component consistent with the balance assumption, $f = nz/\bar{u}$, n = frequency, Hz, z = height, $\bar{u}$ = mean wind speed, u_* = friction velocity, $F_{uu}(f)$ = u-component spectrum function, k = von Kármán constant, z/L = Obukhov stability parameter and $\phi_m = kzu_*^{-1}\, \partial u/\partial z$. The notation is that of Hicks and Dyer (1972) who give the theoretical background in presenting the results of rather similar experiments to those described here, but with less responsive sensing elements. They adopted $k = 0.41$ and this value is used here. It is not significantly different from the value of 0.40 ± 0.01 found by Högström (1985) from experiments over three different surfaces for which the roughness parameters were 1, 7.5, and 24 mm.

2. Experimental Details

The response needed for the sensing elements is dictated by the requirement to measure turbulent intensity in the inertial subrange. Experiments by McIlroy (Deacon, 1959) had shown that, under neutral and unstable conditions, no appreciable transfer of heat and momentum occurs at values of $f > 1$. Hot-wire anemometry was accordingly chosen using vertical platinum wires of diameter 0.012 mm and length 15 mm. These should give negligible attenuation below 10 Hz, so it was arranged to measure the power in the u-spectrum between 5 and 10 Hz and to operate at a height of 4 m. As 10 m s^{-1} was the highest wind speed in these experiments, it follows that all the results are for f-values ≥ 2. This should be satisfactory as many studies show that, under the neutral or unstable conditions dealt with here, the proportionality of spectral intensity to the $-\frac{5}{3}$ power of frequency, which holds in the inertial subrange, extends towards low frequencies as far as $f \approx 0.5$ (see e.g., Kaimal *et al.*, 1972).

The hot-wire anemometer was linearized by the method of Deacon and Samuel (1957) and the output was put through a filter with half-power points at 5 and 10 Hz. This was followed by an amplifier, a 50-cycle chopper, a power amplifier and finally a standard electricity supply watt-hour meter so connected as to square and integrate. The meter, designed for a 50-cycle supply – hence, the chopper – had the gear train removed and the revolutions of the rotor disc counted photoelectrically. A valuable feature for checking the overall gain was the ability to switch over to a dummy anemometer in series with a 0.5 ohm resistor, the PD across which was chopped at 7.5 Hz to give a square-wave output. The odd harmonics associated with a precise square-wave are rejected by the filter and only the fundamental passed. This enabled checks to be made several times each day of the experiments.

Six hot-wire elements were provided and calibrated in a wind tunnel before and after the trials. A different one was used each day to minimize contamination effects and to reduce error in the final mean results. The outputs of the anemometers were ≈ 20 mV per m s^{-1} and the changes between calibration less than 5%, except for that used on 24 November which changed by 7%.

Low-inertia cup anemometers at heights of 1, 2, 4, 8.2, and 16 m provided wind profile data; aspirated temperature elements at 1, 2, and 4 m yielded temperature data. Further particulars of these instruments, masts and methods of mounting are given by Swinbank and Dyer (1968) together with views of sites similar to the 1968 site. In addition to the mast anemometers, there were three anemometers dispersed in the vicinity to give wind speed at about 0.5 m above the surface. The mean speed from these was useful for making preliminary estimates of u_*. Also available were eddy covariance measurements of heat flux by the Fluxatron developed by Hicks (1970).

3. Evaluation of u_* and L

The values of u_* needed in Equation (1) are estimated from the wind speeds at 1, 2, 4, and 8.2 m (with a zero-plane displacement of 2 cm) by using Webb's (1982) profile relationships for unstable conditions. It is assumed that the von Kármán constants for heat and momentum are equal ($k = 0.41$) as appears to be true to a fairly close approximation (Dyer and Bradley, 1982). The appropriate expression is:

$$u_z + (u_*/k)\psi(z/L) = (u_*/k)\ln(z/z_0) \tag{2}$$

in which

$$\psi(z/L) = \int_0^{-z/L} \frac{1-\phi_m}{z/L}\,\mathrm{d}(z/L) \quad \text{and} \quad \phi_m = \frac{kz}{u_*}\frac{\partial u}{\partial z}\,.$$

To apply this, it is necessary first to make a preliminary rough estimate of u_* which, in these experiments over a surface of roughness parameter $z_0 \approx 8$ mm is 10% of the 0.5 m wind speed. This with the Fluxatron heat flux value (increased by an estimated 5% to allow for water vapour buoyancy) gives an approximate L. The best fitting straight line to the corrected wind speeds plotted against log z gives improved estimates of u_* and L. These may differ enough from the first to require one, and rarely two, repetitions of the process. A final fitting by least squares gives u_* and L. The r.m.s. difference between observed wind speeds and the fitted profiles amounts to 2.2 cm s^{-1} or about 0.35% of the wind speed.

The values of L based on the measured heat fluxes, denoted L_H, were checked against those calculated from the differences in wind speed and potential temperature between 1 and 4 m by using Figure 16 of Webb (1982), which enables the Richardson number formed with such differences to be converted to the z/L value at 2 m. The values of L_W so obtained are in good agreement as shown by the mean ratio from 53 runs which is $L_W/L_H = 1.04 \pm 0.025$. Allowing for water vapour buoyancy would bring the ratio even closer to unity.

As Webb's (1982) wind profile formulae are somewhat complicated, a short table of $\psi(z/L)$ and of the stability factor in Equation (1), is given in Appendix 1. Also given there, for comparison, are values according to the formulation of Dyer and Bradley (1982). These authors based their formulae on a different set of observations from those used by Webb.

4. Results and Discussion

Sixty half-hour runs were made covering the range of instability from neutral to $z/L = -0.5$. The 4 m wind speeds ranged between 3.2 and 9.2 m s^{-1}. Figure 1 shows

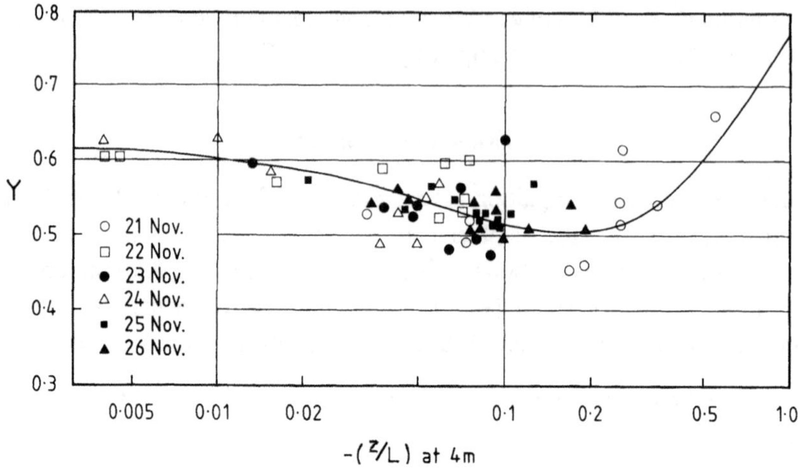

Fig. 1. Variation with stability of $Y = (2\pi k)^{2/3} f^{5/3} F_{uu}(f) u_*^{-2}$ (i.e., right-hand side of Equation (1) omitting the stability factor). The full line is the variation expected according to Webb's (1982) stability factor as given in Appendix 1, Table II.

values of α_{UB} as from Equation (1) but without inclusion of the stability factor in the denominator. The full line is the expected variation according to Webb's (1982) expression for ϕ_m in terms of z/L. There is satisfactory agreement and the results for the several days mesh quite well. After inclusion of the stability factor, the result is:

$$\text{Mean } \alpha_{UB} = 0.625 \pm 0.006.$$

The scatter of results is satisfactorily small as indicated by the standard deviation of values from the mean of 0.042, so that the standard error of the mean of the 60 observations is only 1%. Systematic error is undoubtedly greater than this, but is difficult to estimate accurately: it could well amount to 5% or rather more but is almost certainly no greater than 10%.

These results suggest that, under the conditions of these experiments, the ratio of the rates of dissipation to production of turbulent kinetic energy by shear and buoyancy, if not unity, is at least a constant over the range from neutral stability as far as $z/L = -0.5$. This was also found by Hicks and Dyer (1972) from observations made under similar conditions as to site and observation height (4 m), but their observations show constancy even as far as $z/L = -1.4$.

Some other studies have indicated otherwise. McBean *et al.* (1971) used a sonic anemometer for the inertial subrange and a hot-wire anemometer to enable dissipation to be deduced from the dissipation range of the spectrum. Observations at a height of 2 m over a mown grass surface gave $\alpha_U = 0.54 \pm 0.04$ and balance as per Equation (1) for slightly unstable conditions but dissipation exceeded production by some 20% for z/L around -0.5.

A possibly greater departure from balance is indicated by the results of Wyngaard and Coté (1971) who also used sonic and hot-wire anemometers, but at three heights, 5.7,

11.3, and 22.6 m over a large area of wheat stubble. They found dissipation to exceed production of turbulent K.E. by 30 to 40% over most of the range of z/L from -0.1 to -0.8. However, they point out that had they used $\alpha_U = 0.54$ as given by matching the hot-wire results in the inertial subrange to those from the sonic anemometers, the dissipation rates would come out 15% smaller so that at $z/L = -0.5$ the excess of dissipation over production would again be about 20%.

Over the sea during BOMEX, Pond *et al.* (1971) used a sonic anemometer at a height of 8 m and assuming $\alpha_U = 0.55$, found dissipation to exceed production by amounts increasing with instability over the range of z/L from -0.1 to -0.33. However, the authors estimate that their results have an uncertainty of 15 to 25% owing to effects of disturbance to the wind flow caused by the considerable bulk of the floating instrument platform R/V Flip.

TABLE I

Values of α_{UB} from various researches

Reference and surface type	Anemometer and height, m	No. of obsns.	Wind speed range, m s^{-1}	Range of z/L	Range of nz/u	α_{UB}
Miyake *et al.*, 1970 sea near shore	sonic 1.4–4.5	8	4–9.5	−0.17/0	2–5	0.63 ± 0.04
Pond *et al.*, 1971 sea (Flip)	sonic 8	15	4–7.5	−0.33/−0.1	0.5–10	0.77 ± 0.04
Hicks and Dyer, 1972 grassland	thermistor 4	37	3–7	−1.0/−0.1	0.15–0.5	0.57 ± 0.03
Garratt, 1972 lake	mini-cup 4 and 12	19	4–9	−0.4/0	0.3–3	0.50 ± 0.02
Leavitt, 1975 sea (Flip)	sonic 8 and 30	36	4–8.5	−1.3/−0.1	2–100	0.58 ± 0.02
Wucknitz, 1976 sea (buoy)	hot-wire 2.4	26	5–8.5	−0.03/0	2–10	0.52 ± 0.04
Large and Pond, 1981 sea (spar buoy)	Gill prop. 12.5	186	5–18	−0.6/0.1	0.5–4	0.59 ± 0.015
Dyer and Hicks, 1982 grassland	sonic 4	19	4–8	−0.4/0.1	1–3	0.59 ± 0.01
Frenzen and Hart, 1983 grassland	mini-cup 3 to 8	4	11–13	−0.35/0	2–3.5	0.58 ± 0.02
This report grassland	hot-wire 4	60	3–9	−0.3/0	2–6	0.62 ± 0.01

Some notes on the above are given in Appendix 1.

Table I lists particulars of studies made embracing measurements of momentum flux by eddy covariance along with measurements of spectral intensity in the inertial subrange or close to it. The last column gives the values of α_{UB} as per Equation (1).

The six studies made over water, under mainly slight to moderate instability, give values of α_{UB} ranging from 0.50 to 0.77 with a mean of 0.60 ± 0.04. The rather large scatter is probably due, in part, to the difficulty of securing suitable exposure for the instruments needed for the exacting stress measurements. There may also be a contribution from interaction between turbulence and the waves. Schmitt *et al.* (1978) find evidence for a departure from isotropy of the small scale eddy motion in the lowest few metres over the sea.

Over grassland, the four sets of observations covering a range of unstable conditions give a mean $\alpha_{UB} = 0.59 \pm 0.01$. The means for land and sea are not significantly different. All ten results together give $\alpha_{UB} = 0.59 \pm 0.025$.

The question now arises as to whether this mean value exceeds the true value of α_U in the inertial subrange relationship, as might be expected from the studies already mentioned, which found that dissipation exceeded production of turbulent K.E. under unstable conditions. A difficulty here is that the true value of α_U is not known with much precision. Large and Pond (1981) take $\alpha_U = 0.55 \pm 10\%$ as being reasonable in the light of present knowledge. It seems possible to say only that α_{UB} may well exceed α_U, but probably by no more than about 15% under conditions of moderate instability.

Under strong wind conditions at sea, such as make reliable observations of drag by the eddy covariance method extremely difficult, it is to be expected that the consequent near neutral conditions should favour the spectral density method. That it gives acceptable accuracy was shown by Large and Pond (1981) by comparing its results with simultaneous covariance values over the sea in 10 m winds ranging between 6 and 18 m s^{-1}. Furthermore, their values of the drag coefficient by the spectral density method are less than 10% different from the eddy covariance values obtained at the same site by Smith (1980) using distinctly different instrumentation. These agreements enabled Large and Pond (1981) to employ the spectral density technique on a ship and so determine the trend of the drag coefficient at wind speeds up to 26 m s^{-1}, which is a whole gale, Beaufort force 10.

Earlier shipboard spectral density observations by Denman and Miyake (1973) gave a drag coefficient increasing more slowly with wind speed than found by either Smith or Large and Pond, but converging towards their values at the highest wind speeds.

Finally, it is interesting to recall that Batchelor (1950) in drawing attention to the Kolmogoroff hypotheses about the small-scale structure of turbulent motion, posed the question 'are there any meteorological problems in which the similarity theory would provide useful information?' The successful application of inertial subrange theory to the oceanic drag problem is perhaps a rather unexpected outcome.

Acknowledgements

The observations reported here were secured at what is now known as the Division of Atmospheric Research, CSIRO, and the author wishes to thank the Chief of Division for providing facilities for completion of the analysis.

The electronics for this study were designed and constructed by Mr F. J. Maher, who also contributed the method for squaring and integrating in one operation as well as the square-wave checking system.

References

Antonia, R. A., Chambers, A. J., Rajagopalan, S., and Sreenivarsan, K. R.: 1978, 'Measurements of Turbulent Fluxes in Bass Strait', *J. Phys. Oceanog.* **8**, 28–37.

Batchelor, G. K.: 1950, 'The Application of the Similarity Theory of Turbulence to Atmospheric Diffusion', *Quart. J. Roy. Meteorol. Soc.* **76**, 133–146.

Deacon, E. L.: 1959, 'The Measurement of Turbulent Transfer in the Lower Atmosphere', *Adv. Geophys.* **6**, 211–218.

Deacon, E. L. and Samuel, D. R.: 1957, 'A Linear Temperature Compensated Hot-Wire Anemometer', *J. Sci. Instrum.* **34**, 24–26.

Denman, K. L. and Miyake, M.: 1973, 'Behaviour of the Mean Wind, the Drag Coefficient and the Wave Field in the Open Ocean', *J. Geophys. Res.* **78**, 1917–1931.

Dyer, A. J. and Bradley, E. F.: 1982, 'An Alternative Analysis of Flux-Gradient Relationships at the 1976 ITCE', *Boundary-Layer Meteorol.* **22**, 3–19.

Dyer, A. J. and Hicks, B. B.: 1982, 'Kolmogoroff Constants at the 1976 ITCE', *Boundary-Layer Meteorol.* **22**, 137–150.

Frenzen, P. and Hart, R. L.: 1983, 'A Further Note on the Kolmogorov–von Kármán Product and the Values of the Constants', *Sixth Symposium on Turbulence and Diffusion*, Amer. Meteorol. Soc., Boston, 24–27.

Garratt, J. R.: 1972, 'Studies of Turbulence in the Surface Layer over Water (Lough Neagh), Part II: Production and Dissipation of Temperature and Velocity Fluctuations', *Quart. J. Roy. Meteorol. Soc.* **98**, 642–657.

Garratt, J. R.: 1975, 'Limitations of the Eddy-Correlation Technique for the Determination of Turbulent Fluxes near the Surface', *Boundary-Layer Meteorol.* **8**, 255–259.

Hicks, B. B. and Dyer, A. J.: 1972, 'The Spectral Density Technique for the Determination of Eddy Fluxes', *Quart. J. Roy. Meteorol. Soc.* **98**, 838–844.

Hicks, B. B.: 1970, 'The Measurement of Atmospheric Fluxes near the Surface: A Generalized Approach', *J. Appl. Meteorol.* **9**, 386–388.

Högström, U.: 1985, 'Von Kármán's Constant in Atmospheric Boundary Layer Flow: Reevaluated', *J. Atmos. Sci.* **42**, 263–270.

Kaimal, J. C., Wyngaard, J. C., Izumi, Y., and Coté, O. R.: 1972, 'Spectral Characteristics of Surface-Layer Turbulence', *Quart. J. Roy. Meteorol. Soc.* **98**, 563–589.

Krügermeyer, L.: 1976, 'Vertical Transports of Momentum, Sensible and Latent Heat from Profiles at the Tropical Atlantic During ATEX', *'Meteor' Forsch.-Ergbenisse*, B, No. 11, 51–77.

Large, W. G. and Pond, S.: 1981, 'Open Ocean Momentum Measurements in Moderate to Strong Winds', *J. Phys. Oceanog.* **11**, 324–336.

Leavitt, E.: 1975, 'Spectral Characteristics of Surface Layer Turbulence over the Tropical Ocean', *J. Phys. Oceanog.* **5**, 157–163.

McBean, G., Stewart, R. W., and Miyake, M.: 1971, 'The Turbulent Energy Budget near the Surface', *J. Geophys. Res.* **76**, 6540–6549.

Miyake, M., Stewart, R. W., and Burling, R. W.: 1970, 'Spectra and Cospectra of Turbulence over Water', *Quart. J. Roy. Meteorol. Soc.* **96**, 138–143.

Pond, S., Phelps, G. T., Paquin, J. E., McBean, G., and Stewart, R. W.: 1971, 'Measurements of the Turbulent Fluxes of Momentum, Moisture, and Sensible Heat over the Ocean', *J. Atmos. Sci.* **28**, 901–917.

Schmitt, K. F., Friehe, C. A., and Gibson, C. H.: 1978, 'Sea Surface Stress Measurements', *Boundary-Layer Meteorol.* **15**, 215–228.

Smith, S. D.: 1980, 'Wind Stress and Heat Flux over the Ocean in Gale Force Winds', *J. Phys. Oceanog.* **10**, 709–726.

Swinbank, W. C. and Dyer, A. J.: 1968, 'Micrometeorological Expeditions 1962–1964', *CSIRO Aust. Div. Meteorol. Phys. Tech. Pap.*, No. 17.

Webb, E. K.: 1982, 'Profile Relationships in the Superadiabatic Surface Layer', *Quart. J. Roy. Meteorol. Soc.* **108**, 661–688.

Wucknitz, J.: 1976, 'Determination of Turbulent Fluxes of Momentum and Sensible Heat from Fluctuation Measurements and the Structure of the Wind Field over the Waves above the Tropical Atlantic Ocean during ATEX', *'Meteor' Forsch-Ergebnisse*, B, No. 11, 25–50.

Wyngaard, J. C. and Coté, O. R.: 1971, 'The Budget of Turbulent Kinetic Energy and Temperature Variance in the Atmospheric Surface Layer', *J. Atmos. Sci.* **28**, 190–201.

Appendix 1. The Stability Factor in Equation (1) and Notes on Table I

The stability factor

The values of the stability factor $S(z/L) = (\phi_m - z/L)^{-2/3}$ in Equation (1) are given in Table II as calculated from the formulae of Webb (1982) and of Dyer and Bradley (1982), assuming the von Kármán constants for momentum and heat transfer to be 0.41. Also given are the values of $\psi_m(z/L)$ needed in the evaluation of the shear stress from the wind profiles.

TABLE II

Values of the stability factor $S(z/L)$ and $\psi_m(z/L)$

$-z/L$	0.01	0.02	0.05	0.1	0.2	0.5	1.0	2.0
$S(z/L)$ Webb	1.02	1.05	1.13	1.21	1.23	1.03	0.81	0.58
$S(z/L)$ D and B	1.03	1.06	1.11	1.15	1.14	1.00	0.79	0.56
$\psi_m(z/L)$ Webb	0.045	0.09	0.22	0.42	0.70	1.19	1.59	2.05
$\psi_m(z/L)$ D and B	0.065	0.12	0.26	0.42	0.65	1.05	1.42	1.84

Notes on table I

The values of α_{UB} given in Table I have been derived after making some adjustments to the published data in some cases. Where the authors have assumed a value of the von Kármán constant differing from 0.41, correction has been made to accord with this value. Then in Equation (1) a number of different forms of the stability factor have been used by various authors. Wherever possible, adjustment has been made to give values of α_{UB} corresponding to Webb's values as given in Table II.

In some cases, values of wind gradient $\partial u/\partial z$ were available from wind profiles in which case a more direct evaluation of the rate of production of turbulent energy gives:

$$\alpha_{UB} = fF_{uu}(f)\left\{\frac{2\pi}{z}\right\}^{2/3}\left(u_*^2\,\partial u/\partial z - \frac{gH_+}{\rho c_p}\right)^{-2/3} \tag{2}$$

in which H_+ is the sensible heat flux plus 7% of the latent heat flux to allow for its buoyancy contribution.

Miyake et al. (1970). The data given by these authors along with those in the adjacent companion paper, enable Equation (2) to be used, so avoiding assumptions as to the form of ϕ_m.

Pond et al. (1971). Some uncertainty in the eddy covariance values of u_* in this study was due to complications in analysis resulting from deflection of the wind stream caused by the bulk of the floating instrument platform 'Flip'.

Hicks and Dyer (1972). The spherical bead thermistor anemometer was not responsive enough to sense velocity fluctuations in the inertial subrange proper and the resulting

α-value is only given half weight in arriving at the Table I mean. The results for $z/L < -1$ have been excluded.

Leavitt (*1975*). These data were also secured on 'Flip', so similar remarks to those for Pond *et al.* (1971) apply.

Wucknitz (*1976*). To adjust these data for Webb's stability factor, the values of z/L were obtained with the heat fluxes and latent heat fluxes given by Krügermeyer (1976) in his Table III.

Large and Pond (*1981*). These authors assumed $\phi_m = (1 - 16z/L)^{-1/4}$ for unstable conditions. Insufficient data are given to enable correction to be made. With Webb's form, the value of α_{UB} would be somewhat larger – probably by nearly 10%.

Dyer and Hicks (*1982*). These authors used the stability factor of Dyer and Bradley. Webb's factor would have given $\alpha_{UB} = 0.63$. However as the relationship used by Dyer and Hicks had been derived from the wind profiles observed during the same International Turbulence Comparison Experiment of 1976, no adjustment has been made to their published value.

Frenzen and Hart (*1983*). All the experimental values between $f = 2$ and 3.5 shown in these authors' Figure 1 were used and those for 1976 corrected for the small instability of z/L at 5 m of about -0.033.

The results of some studies have not been included for reasons given in the following.

Antonia et al. (*1978*). In these experiments, a Gill propeller array at 5 m above the sea surface was used to give u_*. That the Gill gear underestimates fluxes when only 5 m above the sea has been shown by Garratt (1975). That this is the case is supported by the low drag coefficient found. Referred to 10 m and neutral conditions, it was 0.93×10^{-3}. This is some 20% less than expected from other researches. So the $\alpha_{UB} \approx 0.75$ indicated by the data in this paper is likely to be too great by a similar amount.

Schmitt et al. (*1978*). In this study, considerable pains were taken to secure a good exposure over the sea for the sensing elements. However, the results of their six observations in winds of 6 to 7 m s^{-1} are too scattered to give a useful mean α_{UB}. Furthermore, the covariance stress values give a suspiciously low drag coefficient of $c_{10} = 0.76 \times 10^{-3}$, which is less than for an aerodynamically smooth surface.

FLOW-DISTORTION EFFECTS ON SCALAR FLUX MEASUREMENTS IN THE SURFACE LAYER: IMPLICATIONS FOR SENSOR DESIGN

J. C. WYNGAARD

National Center for Atmospheric Research, Boulder, CO 80307, U.S.A.*

(Received in final form 1 December, 1986)

Abstract. Scalar fluxes measured through the eddy-correlation technique are prone to two types of errors caused by the sensor-induced flow distortion: those due to crosstalk from the horizontal flux, and those due to amplification or attenuation due to flow blocking. We show that the crosstalk error can be eliminated by designing the sensor array to be vertically symmetric about its horizontal midplane. In such an array, the flow-blocking effect causes the scalar flux to be overestimated, but this error can be made negligible by designing an array with minimal stagnation loss in streamwise speed at the flux-measurement point.

1. Introduction

While there is a long history of research into the effects of flow distortion on mean wind measurements (see, for example, Cermak and Horn, 1968; Izumi and Barad, 1970), much less attention has been paid to the effects on turbulence statistics. This situation began to change in 1980, however, in part due to Wieringa's (1980) assertion that certain of the turbulence data from the 1968 Kansas experiments (Businger *et al.*, 1971; Haugen *et al.*, 1971) had significant errors due to the flow distortion caused by the tower structure. In disputing this, Wyngaard *et al.* (1982) showed that Wieringa's technique for calculating the effects of flow distortion on turbulence was only approximate; they argued that it substantially overestimated the errors. All agreed (Wieringa, 1982) that further analysis of the Kansas data was not likely to shed new light. Nonetheless, Wieringa succeeded in focusing attention on a long-neglected problem in micrometeorology.

Stimulated by Wieringa's study, I presented a simple theory (Wyngaard, 1981) for the effects of flow distortion on turbulence when the scale of the distorting body is small compared to the integral scale of the turbulence. This is almost always the case with distortion by sensors measuring turbulence statistics in the energy-containing range in the surface layer. That first paper concentrated on velocity variance and covariance measurements ahead of a sphere and a cylinder; it also showed that flow-distortion effects on turbulence could not be reliably removed by treating them as 'tilt' errors, as was done by Wieringa (1980). A second study (Wyngaard *et al.*, 1985) extended these results to axisymmetric bodies.

Researchers are now using the eddy-correlation technique to measure vertical fluxes of scalars other than temperature, including O_3 (Wesely *et al.*, 1978), NO_2 (Wesely *et al.*,

* The National Center for Atmospheric Research is sponsored by the National Science Foundation.

1982), CO_2 and water vapor (Ohtaki and Matsui, 1982), aerosols (Fairall, 1984), and NO (Delany *et al.*, 1986). Efforts are underway to develop fast-response sensors for other trace constituents as well. These eddy-correlation devices are typically less than aerodynamically 'clean', however, and the resulting flow distortion has concerned a number of researchers. Current practice is to remove flow-distortion errors in turbulence statistics through use of 'tilt corrections' (e.g., Dyer *et al.*, 1982), although recent lively discussion (Wyngaard *et al.*, 1982; Wieringa, 1982; Dyer, 1982) reveals an emerging consensus that this is *ad hoc* and that flow-distortion corrections should be more fundamentally based.

In concluding their discussion on flow distortion effects in an international turbulence comparison experiment, Dyer *et al.* (1982) write "... it seems abundantly clear that in some cases the sensors themselves would have introduced significant distortion of the flow to an extent that may not be readily predictable, and considerable care must be taken in the basic design of turbulence sensors." In order to provide guidelines for that design process, I will show what flow-distortion theory suggests about the optimum geometry of instruments for the measurement of scalar fluxes near the surface.

2. Scalar Flux Measurement Errors

We denote the fluctuations in scalar mixing ratio and vertical velocity by c and u_3, respectively, and assume that they are measured by sensors located sufficiently close together and of sufficiently small path length or averaging volume. In general, this means the separation and path length are small compared to the height above the surface; we discuss more specific criteria later. We assume that the tower structure and booms have negligible effects on the flow at the sensors. Nevertheless, the measured u_3 signal will be somewhat in error because of the flow distortion caused by the sensors themselves, and this causes an error in the measured scalar flux.

In analyzing this problem, we assume that the sensor housings and mounting apparatus are small compared to the integral length scale of the u_3 field. Since this integral scale is of the order of the height above the surface (Kaimal *et al.*, 1972), this amounts to requiring that the instrument scale is an order of magnitude smaller than the measurement height. This makes the underlying flow-distortion theory simpler and also allows us to ignore the vertical variations of the mean wind and mean concentration (Wyngaard, 1981).

Under these conditions in the surface layer, the flow distortion not only attenuates or amplifies the vertical velocity, but also contaminates it with the horizontal velocity fluctuations. Thus, we can write the vertical velocity fluctuation measured in the region of flow distortion as (Wyngaard, 1981)

$$u_3^m = (1 + d_{33})\, u_3 + d_{31} u_1 + d_{32} u_2 \,. \tag{1}$$

The d_{ij} are small coefficients that approach zero in the distortion-free region far from the body. Under the given assumptions, they can be calculated from the solution for potential flow approaching the body at an arbitrary angle (Wyngaard, 1981). Here d_{33}

represents attenuation or amplification, depending on its sign, while the off-diagonal coefficients d_{31} and d_{32} represent crosstalk and have an equally simple physical interpretation. As a unidirectional approach flow nears a three-dimensional body, the blocking effect induces velocity components in the other two directions. The same effect occurs for large-scale turbulent eddies, and in this way u_1 and u_2 fluctuations induce u_3 fluctuations near the body, as indicated in (1).

Equation (1) assumes that u_3 is measured at a point, but in practice it is usually averaged over a path or a volume, depending on the type of anemometer. This creates no mathematical difficulty, however; by assumption, the only spatial dependence in (1) is in the d_{ij}, so we can formally average (1) over the sensing volume and interpret the d_{ij} as spatially averaged quantities.

We assume that the scalar sensor is of the open-path variety without devices such as aspirating pumps that can lead to wind-gust contamination of the scalar fluctuations. Particle measurements, for example, are subject to such errors (Wesely and Hicks, 1979; Fairall, 1984). Since the local Peclet number of the flux-carrying scalar fluctuations is large, we can neglect the effects of molecular diffusion during the travel time near the sensor. Thus, to a good approximation the total time derivative of the scalar vanishes, and it suffers no changes due to flow distortion:

$$c^m = c\,. \tag{2}$$

While this indicates that the amplitude of scalar fluctuations is unchanged by flow distortion, their spatial scale clearly can be changed. These deformation effects can be very important when measuring scalar fine structure (Wyngaard, 1986) but should be of no consequence for flux measurements provided that the flux-carrying eddy scale is large compared to the probe, as we have assumed.

Multiplying (1) and (2) and averaging gives an expression for the measured flux:

$$\overline{u_3^m c^m} = (1 + d_{33})\,\overline{u_3 c} + d_{31}\,\overline{u_1 c} + d_{32}\,\overline{u_2 c}\,. \tag{3}$$

If we choose u_1 to be in the mean wind direction, then by lateral symmetry we would expect that $\overline{u_2 c} = 0$ under horizontally homogeneous conditions; this is confirmed by observations of temperature fluxes (Zubkovskii and Tsvang, 1966). However, the streamwise flux $\overline{u_1 c}$ does not vanish. This was apparently first demonstrated experimentally for temperature by Shiotani (1955) and confirmed by Zubkovskii and Tsvang (1966). Wyngaard *et al.* (1971) extended these results for temperature to a wide range of stabilities, and showed that the horizontal scalar flux is produced through the interaction of turbulence with the vertical gradients of the mean scalar concentration and mean wind and is of opposite sign to the vertical flux.

Although the Wyngaard *et al.* (1971) analysis was for fluxes of temperature, their findings should hold equally well for any conservative scalar in the surface layer, and so I present their results as $\overline{u_1 c}/\overline{u_3 c}$ in Figure 1. Note that in stable and near-neutral conditions $\overline{u_1 c}/\overline{u_3 c} \sim -3$, so that (3) yields

$$\overline{u_3^m c^m}/\overline{u_3 c} \simeq 1 + d_{33} - 3d_{31}\,. \tag{4}$$

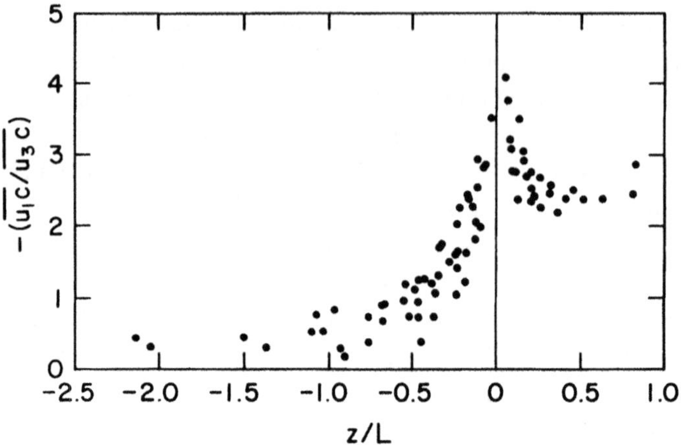

Fig. 1. The ratio of streamwise and vertical components of scalar flux in the surface layer, as measured for temperature in the 1968 Kansas experiments. Data taken at 5.66, 11.3, and 22.6 m and plotted against the stability index z/L. From Wyngaard *et al.* (1971).

This suggests that the effect of crosstalk is potentially more damaging than that of the attenuation/amplification represented by the d_{33} term in (4).

3. Implications for Sensor Design

The d_{ij} are formally defined (Wyngaard, 1981) through the coefficients in a Taylor series expansion of the distorted flow field about a basic state of uniform mean approach flow of velocity $(U_1, 0, 0)$, The crosstalk coefficient d_{31} in (4) is

$$d_{31}(x) = \left.\frac{\partial \tilde{U}_3(\mathbf{x})}{\partial U_1}\right|_0, \tag{5}$$

where the tilde represents the distorted flow and U_1 is the free-stream speed. While we can calculate d_{31} for simple bodies and in general we can measure it experimentally, in this case we can also write a simple approximation for it. In the surface layer, we can take the undistorted mean vertical velocity U_3 to be zero, so that for small d_{31} we can approximate (5) by

$$d_{31} \simeq \tilde{U}_3/U_1 \simeq \sin\theta, \tag{6}$$

which is simply the mean deflection angle of the airflow at the measurement point. This allows us to estimate the flux error caused by crosstalk. If we have 5 deg of flow deflection, for example, which seems quite possible in compact, vertically asymmetric arrays, then $d_{31} \sim 0.1$, and (4) indicates that the flux error under near-neutral conditions is about 25%. Note that the flux is overestimated if d_{31} is negative, which occurs for 'top-heavy' geometries, with the reverse true for the 'bottom-heavy' case.

We can now compare the distorted u_3 signal (1) with that caused by a tilt error – that is, by vertical misalignment of an otherwise perfect u_3 sensor. If the tilt angle from true vertical is θ, the measured u_3 signal is

$$u_3^m = u_3 \cos\theta + u_1 \sin\theta. \tag{7}$$

In view of (6), the crosstalk contribution of u_1 to u_3^m due to flow distortion (the second term in (1)) is the same as the crosstalk contribution of u_1 to u_3^m due to tilt error (the second term in (7)). The analogy does not extend to the first terms, however. As we shall soon see, the coefficient $(1 + d_{33})$ of the u_3 term in the flow-distortion equation (1) can be greater or less than 1.0, depending on the geometry of the distortion, whereas that coefficient in the tilt equation (7) is simply the cosine of the tilt angle and cannot exceed 1.0 in magnitude. Thus, the use of 'tilt corrections' for flow-distortion effects is incorrect.

The result (6) highlights an important property of d_{31}: it is identically zero on a horizontal plane of symmetry, because on such a plane $\tilde{U}_3$ is zero, by definition. Our first design criterion, then, is that the sensor array (i.e., the combination of c and u_3 sensors) should be symmetric about the horizontal plane passing through the point of measurement of u_3. This will eliminate the crosstalk error.

The need for vertical symmetry in u_3 sensors seems not to have been stressed in the literature. However, Hicks (1972) discussed the updraft caused by the asymmetry of a propeller anemometer mounted with the axis vertical, indicating that the resulting mean flow deflection in one such application was about 1.2 deg. He recommended the use of a vertical shaft extension beyond the plane of the propeller to minimize the asymmetry. More recently, Kaimal (1986) recommended incorporating 'as much vertical symmetry as possible into the probe design'.

Let us assume, therefore, that we design a vertically symmetric array. According to (4), we must now minimize the blocking error represented by d_{33}. To get some insight into this error, let us assume further that the blocking effect of the array is equivalent to that of a sphere. Calculations of d_{33} for a sphere (Wyngaard, 1981) indicate that it is positive within an approach cone of about 70 deg (Wyngaard, 1981) and negative outside. In a symmetric array, we make the u_3 measurement along the centerline, where d_{33} is

$$d_{33} = \frac{a^3}{2r^3} \tag{8}$$

with a the sphere radius and r the distance between the sphere center and the point where u_3 is measured. Thus, we see that for a symmetric array, u_3 is amplified by the flow distortion, leading to an overestimate of scalar flux. This amplification (which is ignored by 'tilt corrections' for flow distortion effects) has been demonstrated in the measurements of Bearman (1972) and Britter *et al.* (1979) ahead of a circular cylinder in turbulent flow.

Potential flow theory says that the decrease in streamwise speed along the centerline ahead of the sphere is

$$1 - \frac{\tilde{U}_1}{U_1} = \frac{a^3}{2r^3}. \tag{9}$$

If we combine (8) and (9), we can eliminate the sphere radius a and relate d_{33} simply to the mean-flow stagnation effect:

$$d_{33} \simeq 1 - \frac{\tilde{U}_1}{U_1} . \tag{10}$$

This result also holds exactly for a circular cylinder (Bearman, 1972; Wyngaard, 1981) and, judging from the numerical calculations of Wyngaard *et al.* (1985), is a good approximation for an axisymmetric body. Thus, we conclude that a reasonable estimate of the ratio of true and measured scalar fluxes in a symmetric array is, from (4) and (10),

$$\frac{\overline{u_3^m c^m}}{\overline{u_3 c}} \simeq 2 - \frac{\tilde{U}_1}{U_1} . \tag{11}$$

Our estimate of the fractional error in measured scalar flux due to blocking effects is simply the negative of the fractional change in streamwise speed, at the u_3 measurement point, due to stagnation.

Our second design criterion, then, is to minimize the bulk of the array so that u_3 is measured at a point where the stagnation loss in streamwise speed is minimal.

Meeting this criterion will typically require a separation between the u_3 and c sensors. While calculating the effect of this separation on flux measurements is straightforward in principle, in practice it is difficult because we do not have the detailed information on the flux cospectrum needed to evaluate the resulting integrals (Wyngaard, 1986). Spectral modeling can provide useful guidelines here, however (e.g., Kristensen and Fitzjarrald, 1984). One can also evaluate separation effects experimentally, as Koprov and Sokolov (1973) attempted to do. While not complete enough to serve as the basis for design, their results do indicate that flux loss is more sensitive to lateral separation than to vertical separation. Finally, one can also improvise conservative rules of thumb, as have been suggested for sonic anemometry (e.g., Kaimal, 1986). One such rule would be that sensor separation equal to path length would cause no additional flux degradation, providing the path length itself were sufficiently short; a conservative criterion for the latter can, in turn, be developed by considering the spectral response, which can be evaluated numerically (Wyngaard, 1986). This probably can be relaxed for the separation component in the streamwise direction, since to a good approximation this can be corrected for by lagging the upwind sensor signal in time to account for the transit time difference between probes.

Sonic anemometers are often used in scalar-flux sensors because of their excellent dynamic response (Kaimal, 1986). However, they are prone to flow-distortion errors stemming both from the bulk of the sonic array and from the wakes of the acoustic transducers. The latter, called the 'transducer-shadow effect', can cause substantial errors in the spectral response (Wyngaard and Zhang, 1985). The flow-distortion errors due to the array bulk have apparently yet to be investigated systematically, but it is encouraging that recent designs discussed by Wyngaard and Zhang (1985) and Zhang *et al.* (1986) are vertically symmetric and would seem to cause less blockage than earlier units.

4. Conclusions

Scalar fluxes measured through eddy correlation have two types of errors, each stemming from the effects of probe-induced flow distortion on the measured vertical velocity fluctuations. One has the nature of an instrument tilt error, in that it introduces horizontal velocity fluctuations into the measured vertical ones; this crosstalk effect thereby contaminates the vertical flux with the horizontal one. This is potentially a serious source of error because this horizontal flux is not normally negligible; it exceeds the vertical scalar flux by a factor of three in magnitude in the near-neutral surface layer, for example. The other error is due to attenuation or amplification of the vertical velocity due to flow distortion.

The crosstalk error can be eliminated by designing the eddy-correlation array (i.e., the combination of c and u_3 sensors, housings, and mounts) to be vertically symmetric about the horizontal midplane of u_3 and c measurement.

In such an array the flow-blockage effect amplifies the u_3 signal and thereby causes the scalar flux to be overestimated. We show that to a plausible approximation the fractional error in flux is simply the negative of the fractional loss in streamwise speed at the u_3 measurement point and, hence, can be minimized by designing an array with minimum flow obstruction.

The remaining source of error in this 'optimum array' is that due to path averaging and sensor separation. Given the absence of experimental data on the latter, and the lack of the cospectral information needed to evaluate it analytically, we suggest that the lateral separation be limited to distances on the order of the path length. Streamwise separation can be somewhat larger provided that time-lag adjustments be made to account for the transit-time difference between the two sensors.

Acknowledgments

I am grateful to J. Businger and W. Dabberdt for inviting me to speak on the subject of measurement physics at the STEM (Surface Transfer and Exchange Measurement) Workshop at NCAR and thereby stimulating this note, and for comments on its first draft; to B. Hicks, J. C. Kaimal, D. Lenschow, and anonymous reviewers for their comments; and to M. Farnsworth for preparing the manuscript.

References

Bearman, P. W.: 1972, 'Some Measurements of the Distortion of Turbulence Approaching a Two-Dimensional Bluff Body', *J. Fluid Mech.* **53**, 451–467.

Britter, R. E., Hunt, J. C. R., and Mumford, J. C.: 1979, 'The Distortion of Turbulence by a Circular Cylinder', *J. Fluid Mech.* **92**, 269–301.

Businger, J. A., Wyngaard, J. C., Izumi, Y., and Bradley, E. F.: 1971, 'Flux-Profile Relationships in the Atmospheric Surface Layer', *J. Atmos. Sci.* **28**, 181–189.

Cermak, J. E. and Horn, J. D.: 1968, 'Tower Shadow Effect', *J. Geophys. Res.* **73**, 1869–1876.

Delany, A. C., Fitzjarrald, D. R., Lenschow, D. H., Pearson, Jr., R., Wendel, G. J., and Woodruff, B.: 1986, 'Direct Measurements of Nitrogen Oxides and Ozone Fluxes over Grassland', *J. Atmos. Chem.* **4**, 429–444.

Dyer, A. J.: 1982, 'Reply', *Boundary-Layer Meteorol.* **22**, 267–268.
Dyer, A. J., Garratt, J. R., Francey, R. J., McIlroy, I. C., Bacon, N. E., Hyson, P., Bradley, E. F., Denmead, O. T., Tsvang, L. R., Volkov, Y. A., Koprov, B. M., Elagina, L. G., Sahashi, K., Monji, N., Hanafusa, T., and Tsukamoto, O.: 1982, 'An International Turbulence Comparison Experiment (ITCE 1976)', *Boundary-Layer Meteorol.* **24**, 181–209.
Fairall, C.: 1984, 'Interpretation of Eddy-Correlation Measurements of Particulate Deposition and Aerosol Flux', *Atmos. Env.* **18**, 1329–1337.
Haugen, D. A., Kaimal, J. C., and Bradley, E. F.: 1971, 'An Experimental Study of Reynolds Stress and Heat Flux in the Atmospheric Surface Layer', *Quart. J. Roy. Meteorol. Soc.* **97**, 168–180.
Hicks, B. B.: 1972, 'Propeller Anemometers as Sensors of Atmospheric Turbulence', *Boundary-Layer Meteorol.* **3**, 214–228.
Izumi, Y. and Barad, M. L.: 1970, 'Wind Speeds as Measured by Cup and Sonic Anemometers and Influenced by Tower Structure', *J. Appl. Meteorol.* **9**, 851–856.
Kaimal, J. C.: 1986, 'Flux and Profile Measurements from Towers in the Boundary Layer', in D. H. Lenschow (ed.), *Probing the Atmospheric Boundary Layer*, American Meteorological Society, Boston, pp. 19–28.
Kaimal, J. C., Wyngaard, J. C., Izumi, Y., and Coté, O. R.: 1972, 'Spectral Characteristics of Surface-Layer Turbulence', *Quart. J. Roy. Meteorol. Soc.* **98**, 563–589.
Koprov, B. M. and Sokolov, D. Yu.: 1973, 'Spatial Correlation Functions of Velocity and Temperature Components in the Surface Layer of the Atmosphere', *Izv. Atmos. and Oceanic Phys.* **9**, 95–98.
Kristensen, L. and Fitzjarrald, D. R.: 1984, 'The Effect of Line Averaging on Scalar Flux Measurements with a Sonic Anemometer near the Surface', *J. Atmos. Ocean. Tech.* **1**, 138–146.
Ohtaki, E. and Matsui, T.: 1982, 'Infrared Device for Simultaneous Measurement of Fluctuations of Atmospheric Carbon Dioxide and Water Vapor', *Boundary-Layer Meteorol.* **24**, 109–119.
Shiotani, M.: 1955, 'On the Fluctuation of the Temperature and Turbulent Structure near the Ground', *J. Meteorol. Soc. Japan* **33**, 117–123.
Wesely, M. L., Eastman, J. A., Cook, D. R., and Hicks, B. B.: 1978, 'Daytime Variations of Ozone Eddy Fluxes to Maize', *Boundary-Layer Meteorol.* **15**, 361–373.
Wesely, M. L., Eastman, J. A., Stedman, D. H., and Yalvac, E. D.: 1982, 'An Eddy-Correlation Measurement of NO_2 Flux to Vegetation and Comparison to O_3 Flux', *Atmos. Environ.* **16**, 815–820.
Wieringa, J.: 1980, 'A Revaluation of the Kansas Mast Influence on Measurements of Stress and Cup Anemometer Overspeeding', *Boundary-Layer Meteorol.* **18**, 411–430.
Wieringa, J.: 1982, 'Reply', *Boundary-Layer Meteorol.* **22**, 251–255.
Wyngaard, J. C.: 1981, 'The Effects of Probe-Induced Flow Distortion on Atmospheric Turbulence Measurements', *J. Appl. Meteorol.* **20**, 784–794.
Wyngaard, J. C.: 1982, 'Comments on "Flow Distortion by Supporting Structures" by A. J. Dyer', *Boundary-Layer Meteorol.* **22**, 263–265.
Wyngaard, J. C.: 1986, 'Measurement Physics', in D. H. Lenschow (ed.), *Probing the Atmospheric Boundary Layer*, American Meteorological Society, Boston, pp. 5–18.
Wyngaard, J. C. and Zhang, S. F.: 1985, 'Transducer-shadow effects on turbulence spectra measured by sonic anemometers', *J. Atmos. Oceanic Tech.* **2**, 548–558.
Wyngaard, J. C., Businger, J. A., Kaimal, J. C., and Larsen, S. E.: 1982, 'Comments on "A Revaluation of the Kansas Mast Influence on Measurements of Stress and Cup Anemometer Overspeeding"', *Boundary-Layer Meteorol.* **22**, 245–250.
Wyngaard, J. C., Coté, O. R., and Izumi, Y.: 1971, 'Local Free Convection, Similarity, and the Budgets of Shear Stress and Heat Flux', *J. Atmos. Sci.* **28**, 1171–1182.
Wyngaard, J. C., Rockwell, L., and Friehe, C.: 1985, 'Errors in the Measurement of Turbulence Upstream of an Axisymmetric Body', *J. Atmos. Ocean. Tech.* **2**, 605–614.
Zhang, S. F., Wyngaard, J. C., Businger, J. A., and Oncley, S. P.: 1986, 'Response Characteristics of the U.W. Sonic Anemometer', *J. Atmos. Oceanic Tech.* **2**, 315–323.
Zubkovskii, S. and Tsvang, L. R.: 1966, 'Horizontal Turbulent Heat Flow', *Izv. Akad. Nauk SSSR, Ser. Geofiz.* **2**, 1307–1310.

ENERGY BUDGET FOR THE SAHEL SURFACE LAYER DURING THE ECLATS EXPERIMENT

PIERRE DURAND, JEAN-PIERRE FRANGI*, and AIME DRUILHET

Laboratoire d'Aérologie, Université Paul Sabatier, 118 rue de Narbonne, 31062 Toulouse Cedex, France

(Received in final form 20 October, 1986)

Abstract. The measurements obtained during the ECLATS experiment were used in order to determine the surface energy budget of the Sahel region (Niamey, Niger). This expedition was carried out from November 15 to December 10, 1980, during the dry period. Some data were collected by an instrumented aircraft, from which the turbulent fluxes were obtained in the boundary layer around midday; data were also collected at a surface station in order to estimate the surface energy budget continuously by the profile method. The aircraft measurements show the homogeneity of the vertical fluxes over large areas, allowing generalization to the bushy steppe of the Sahel region. The mean diurnal cycle of the energy budget is characterized by high values of ground heat flux and weak values of latent heat flux (deduced from the balance of the energy budget). This cycle is compared with that of the Koorin expedition, performed in similar conditions (tropical savanna in the dry period). We compare the three midday budgets: during Koorin; during ECLATS, at the ground station, and with the aircraft. The important differences that appear in the net radiative flux are explained by the difference in surface albedo.

1. Introduction

The surface energy budget of the Earth can vary considerably from region to region. Very few data are available for the Sahel region, although the climatic, economic and human problems make it worthy of particular attention. Because of its climatic characteristics, this region constitutes a natural laboratory providing exceptional conditions for the study of meteorological phenomena. Some information was provided by Riou *et al.* (1979) on the relation between the albedo, the humidity and the soil surface temperature in semi-arid regions. Climatic data on heat and water balance in West Africa were gathered by Oladipo (1980).

The 'ECLATS'** experiment was carried out in the neighbourhood of Niamey (Niger) (13°30′ N, 2°20′ E), in November and December 1980 (Druilhet and Tinga, 1982). The aim of this experiment was to study the diurnal cycle of radiative and turbulent transfers in the boundary layer. The site and the period were chosen because of the homogeneity of the surface (bushy steppe) and because of the repetitive character of the diurnal variation of meteorological conditions. During the experiment, we observed a high turbidity due to the 'dust haze' (Bertrand *et al.*, 1975).

The main facilities used were:

– a base station in order to measure continuously the ground energy budget (G.E.B.) in the atmospheric surface layer (A.S.L.) and the optical properties of the atmosphere;

* Ecole des Sciences, Université de Niamey, B.P. 10662 Niamey, Niger.

** ECLATS: Etude de la Couche Limite Atmosphérique Tropicale Sèche.

Boundary-Layer Meteorology **42** (1988) 27–42.

– a tethered balloon used for dynamic and radiative studies in the nocturnal boundary layer (Estournel *et al.*, 1986);

– an instrumented aircraft specially used for investigations in the convective boundary layer. The results lead to a statistical analysis of the turbulent properties of the Sahel boundary layer (Durand, 1983; Druilhet and Durand, 1984). Only the lowest altitude measurements are used here.

All the statistical data gathered on the energy budget of the A.S.L. are presented. The G.E.B. station measurements allow establishment of the diurnal variation of the energy budget at one point. The aircraft is an excellent tool for characterization of fluxes above large areas. The flights were performed mainly around midday. We used as a reference the data gathered in the Koorin experiment, which was carried out in conditions close to those of the ECLATS experiment (tropical region, savanna land, dry period). These data allow computation of the ground energy budget. We compared the diurnal cycle of the energy budget for both experiments and analysed more accurately the budget at midday, since we also have the airborne flux measurements at that time.

2. Meteorological Conditions

Meteorological conditions in the Sahel are controlled by the position of the Intertropical Discontinuity (I.T.D.), which separates the humid monsoon air of the southwest from the dry, hot air of the eastern sector (Harmattan). During the experiment, the I.T.D. was situated about 200 km south of Niamey, so that the air mass was hot and dry. Figure 1 shows, for the whole experiment, the downward solar flux received at ground level as well as the mean hodographs of the wind at 00, 06, 12, and 18 GMT. The solar

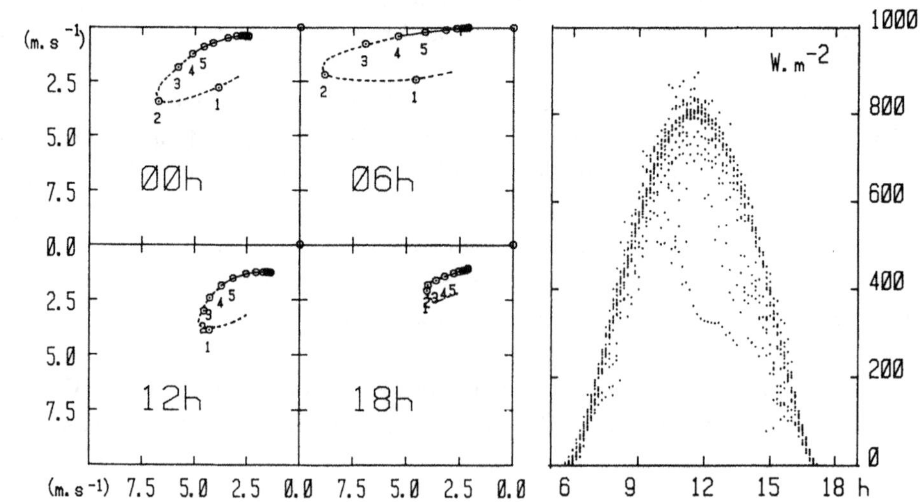

Fig. 1. *On the left*: Wind hodographs (average over the period of the experiment) at 00, 06, 12, 18 (G.M.T.) The numbers 1 to 5 correspond to heights of 100 to 500 m. The highest level is 1500 m. *On the right*: downward solar radiation flux at ground level during the experiment.

flux was measured with an Eppley pyranometer (0.2 to 2.8 μm) and the hodographs were calculated from the wind observations at the meteorological station of Niamey.

The mean solar flux measured at 12 GMT was 800 W m^{-2}, lower values corresponding to the passage of high altitude clouds, mainly at the beginning of the experiment. The variability also depended on the variability of the optical thickness of the atmosphere, i.e., the density of the dust haze. The maximum flux at the top of the atmosphere was 1140 W m^{-2} for the period, the duration of daylight was 11.2 hr and the zenith angle of the sun varied between 28 and 35 deg.

The wind profile at 06 GMT was characterized by a low level jet (9 m s^{-1} at 200 m). During the day the wind decreased and became more uniform in the atmospheric boundary layer, while at 1500 m, the wind was generally weak, and less than 5 m s^{-1}.

The chronological variation of the main meteorological parameters in the A.S.L. is shown in Figure 2 for the whole period: air temperature, relative humidity, vapour

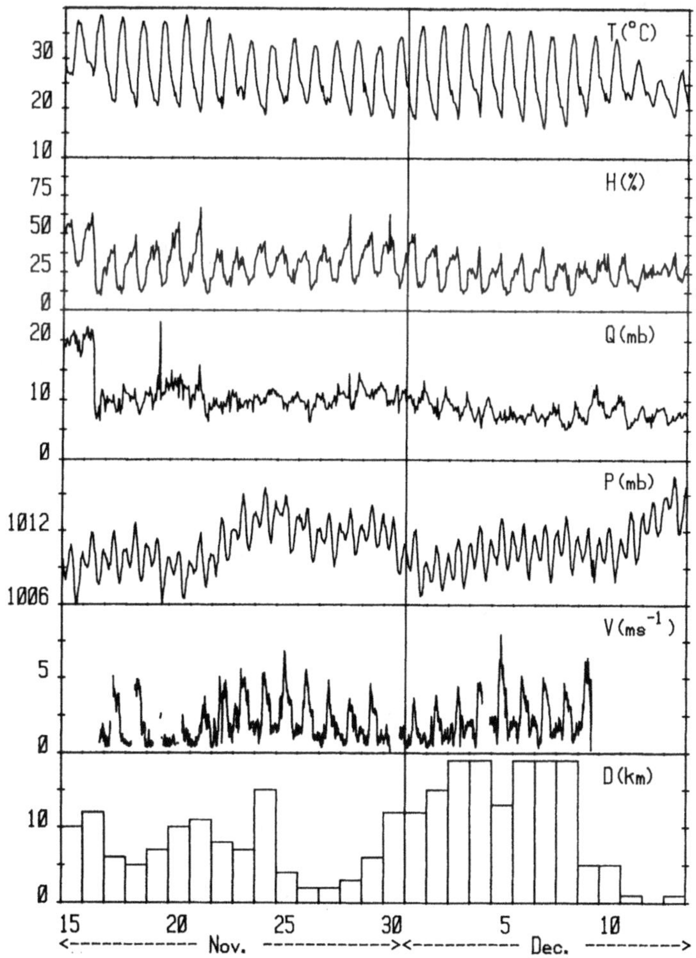

Fig. 2. Variation during the experiment in air temperature (T), relative humidity (H), vapour pressure (Q), atmospheric pressure (P), wind speed at 2.25 m (V), and horizontal visibility (D).

pressure (at 2 m), sea level pressure, wind speed (at 4 m), and horizontal visibility. The wind was very weak at night ($<2\ m\ s^{-1}$). It reached its maximum just after sunrise when thermal instability ensured mixing of the lower layers, and thus a transfer of momentum from the higher levels towards the ground. From 1100 GMT onwards it decreased until the end of the afternoon. This diurnal variation induced a well-defined cycle in the stability of the A.S.L., characterized by a maximum of dynamic instability in the morning. The air temperature reached its minimum at 06 GMT (19 °C) and its maximum at 14 GMT (34 °C). The mean vapour pressure was 10 mb and its diurnal variation was weak, with a minimum during the day. In contrast, the diurnal variation of the relative humidity was high, mainly due to the diurnal variation of the temperature. (35% at the end of the night and 20% in the middle of the day). Diurnal and semi-diurnal pressure waves were also observed. The principal maximum occurs at 1000 GMT and the secondary maximum at 23 GMT. The amplitude of the diurnal wave was 0.75 mb and that of the semi-diurnal wave 1.4 mb. Figure 2 illustrates the coupling between pressure, wind and haze, and the impact of haze on the amplitude of the diurnal cycle.

3. Site and Implementation

3.1. Measurement zones

The G.E.B. station was located near Niamey Airport (13°30′ N, 2°11′ E). This site was at the extreme western end of a plateau (200 m M.S.L.) where the airborne measurements were performed. For a distance of about 1 km around the G.E.B. station, the ground was covered only with thorn bushes (1 to 2 m high) a few tens of meters apart. In the neighbourhood (100 m) of the measurement point, the soil was bare, ochre coloured with a surface composed of fine particles. The immediate subsoil consisted of river sediments (a mixture of gravel and sand with very little clay binder).

The airborne measurements were made over this plateau, situated east of Niamey, over two zones corresponding to squares with 40 km sides. Their SW and NE corners were: 13°10′ N–2°20′ E and 13°30′N–2°40′ E for the first, 13°10′ N–3°00′ E and 13°30′ N–3°20′ E for the second. The boundary conditions of the two zones were similar. The surface comprised the following:

– cultivated areas, generally growing millet. During the measurement period, these areas were bare, with isolated trees, 5 to 10 m high, a few hundred meters apart;

– 'stripped bush', i.e., alternate strips of bare soil and bushes (~2 m high). The orientation of these bands depends on both the slope of the relief and the wind direction. The covered surface did not exceed 30%.

– 'Shrubby steppe' with bushes and isolated trees (up to 10 m high) which covered about 50% of the surface.

Each of these three types of vegetation covered about one third of the aircraft zone.

3.2. Instrumentation and Methodology

3.2.1. *The G.E.B. Station*

The components of the energy budget at ground level are:

Fn: the net radiative flux,
Ft: the sensible heat flux,
Fg: the ground heat flux,
Fl: the latent heat flux.

The net radiative flux was measured with a Swissteco differential pyrradiometer with ventilated polyethylene domes. The instrument was at a height of about 1 m.

The ground heat flux (Fg) was measured using a fluxmeter 0.03 m below the surface. There were two problems with these measurements: the absolute calibration of the sensor; and the correction due to the storage term between the soil surface and 0.03 m depth. These problems were solved by measuring soil temperature profiles down to 1 m during a period of six continuous days, using five differential thermocouples. In order to compute the heat flux, the temperature field was described by a Fourier serie up to fifth order, and the variation with depth of the amplitude and phase of each harmonic was fitted by polynomial functions (third order). The physical characteristics of the soil (heat capacity (C_g), apparent density (ρ_g) and thermal conductivity (K_g)) were also determined:

$$C_g = 1.5 \times 10^6 \text{ J kg}^{-1}\text{ K}^{-1}, \qquad \rho_g = 2.6 \times 10^3 \text{ kg m}^{-3}, \qquad K_g = 7 \times 10^{-7}\text{ m}^2\text{ s}^{-1}$$

We could then deduce the heat flux for any depth, and particularly for the surface and 0.03 m depth. That allowed us to obtain a fluxmeter correction which took into account the calibration, the amplitude attenuation, and phase drift due to the storage in the layer 0–0.03 m.

The sensible heat flux was determined from temperature and wind measurements at two levels, using the universal functions of the A.S.L. The chosen levels were 0.75 and 2.25 m. The wind was measured with cup anemometers; the temperature difference was measured with a differential set of six series of thermocouples. The universal functions ϕ_m and ϕ_T for momentum and heat, respectively, are defined by:

$$\mathrm{d}U/\mathrm{d}z = \frac{U_*}{kz}\,\phi_m\left[\frac{z}{L}\right], \tag{1}$$

$$\mathrm{d}\theta/\mathrm{d}z = \frac{T_*}{kz}\,\phi_T\left[\frac{z}{L}\right], \tag{2}$$

where U is the horizontal wind speed, θ the potential temperature, and k von Karman's constant; U_* and T_* are the friction velocity and the temperature scale respectively. The sensible heat flux is linked to these scales by:

$$Ft = -\rho Cp U_* T_* \tag{3}$$

where ρ is the density and Cp the heat capacity of the air. The integration of (1) and (2) gives:

$$U(z) = \frac{U_*}{k} [\ln(z/z_0) + \Psi_m(z/L)] \tag{4}$$

$$\theta(z) - \theta(z_t) = \frac{T_*}{k} [\ln(z/z_t) + \Psi_T(z/L)] \tag{5}$$

where z_0 is the aerodynamic roughness length and z_t the roughness length for heat. The universal functions used are defined by Dyer and Hicks (1970) and Webb (1970):

- Stable conditions:

$$\phi_m = \phi_T = 1 + 5z/L\,. \tag{6}$$

- Unstable conditions:

$$\phi_m^2 = \phi_T = (1 - 16z/L)^{-0.5}\,, \tag{7}$$

where L is the Monin–Obukhov length:

$$L = \frac{U_*^2 T}{kgT_*}\,. \tag{8}$$

In unstable conditions, the functions Ψ_m and Ψ_T are given by Paulson (1970). Equations (4) and (5) give:

$$U(Z_1) - U(Z_2) = \frac{U_*}{k} [\ln(Z_1/Z_2) + \Psi_m(Z_1/L) - \Psi_m(Z_2/L)]\,, \tag{9}$$

$$\theta(Z_1) - \theta(Z_2) = \frac{T_*}{k} [\ln(Z_1/Z_2) + \Psi_T(Z_1/L) - \Psi_T(Z_2/L)]\,, \tag{10}$$

where Z_1 and Z_2 are the measurement levels. Equations (8), (9), and (10) enable us to calculate L, U_*, and T_*. Further, (4) allows the computation of z_0, whose value for the measurement period was 0.001 m. The latent heat flux (Fl) is estimated from the energy budget balance. This estimation, therefore, accumulates the errors in the calculation of the other three terms. Finally, the soil surface temperature (T_s) was measured by six differential thermocouples located in the first millimeter of the subsoil, and distributed on a circle with a radius of 1 m.

3.2.2. *The Aircraft*

3.2.2.1. *Measured parameters* The aircraft was a Hurel–Dubois 34 of the 'Institut Geographique National' equipped for measuring thermodynamic and turbulence parameters (Durand, 1983), as well as radiation and aerosol properties (Cerf *et al.*, 1982). The airborne measurements were used to estimate:

- the net radiative flux and the albedo of the surface; both these terms were deduced from the four measured components: upward and downward long-wave, and upward and downward short-wave radiative fluxes;

– the soil surface temperature (T_s) deduced from the Barnes PRT5 radiometer, assuming an emissivity of 1;

– the vertical turbulent fluxes:

  – of momentum: $\rho\overline{w'u'}$,

  – of sensible heat: $\rho Cp\overline{w'\theta'}$,

  – of latent heat: $\rho\mathscr{L}\overline{w'q'}$,

where $\mathscr{L}$ is the latent heat of vaporization; w', u', θ', and q' are respectively the fluctuations of the vertical and horizontal speed of the air, the potential temperature and the mixing ratio.

These three fluxes are calculated along horizontal flight sequences of 15 to 20 km, by the eddy correlation method in the wavelength band from 8 m to 2.5 km. The surface fluxes are estimated by extrapolating the fluxes measured at the lowest flight levels. The lowest level is generally between 30 and 50 m.

3.2.2.2. *Spatial homogeneity of fluxes above the steppe* The advantage of airborne measurements relative to the G.E.B. point values is that they are representative of large areas. The methodology of the covariance calculation allows the variability of the flux to be estimated. This estimation is made from the integral function:

$$f_{xy}(t) = \int_0^t x'(t)y'(t)\,dt\,, \tag{11}$$

where the conversion $dl = V_p\,dt$ is made, V_p being the airspeed of the aircraft. The more homogeneous the flux, the closer does this function approach a straight line. A 'statistical' flux can be calculated: it is defined by the slope of the straight line which best fits the function $f_{xy}(t)$. The correlation coefficient R_{xy} of this regression allows the variability of the transfer during the sampling interval to be quantified.

An example of the calculation of turbulent fluxes during a low-level flight (30 m above the ground) is given in Figure 3. The calculated covariances are $\overline{w'\theta'}$, $\overline{w'q'}$, $\overline{w'u'}$. (u' is along the mean wind of the leg.) For each covariance $\overline{w'x'}$, the instantaneous $w'x'$ and the integral curve $f_{xy}(t)$ are also shown. For each flux, these curves are well fitted by a straight line: the correlation coefficients defined above are: $R_{wt} = 0.999$ for the sensible heat flux, $R_{wq} = 0.993$ for the latent heat flux, and $R_{wu} = -0.995$ for the momentum flux.

These high values show the high degree of homogeneity along this path. Figure 4 generalizes this result for all the experiments: it presents the correlation coefficients R_{wt} and R_{wq} as a function of the reduced altitude z/Zi for all the flight sequences performed in the convective boundary layer (Zi is the thickness of the A.B.L.). From the ground to a height of $0.3\,Zi$, most values lie between 0.95 and 1, from which we conclude that the low-level fluxes are very homogeneous and thus representative of the type of surface analysed (steppe in the Sahel region).

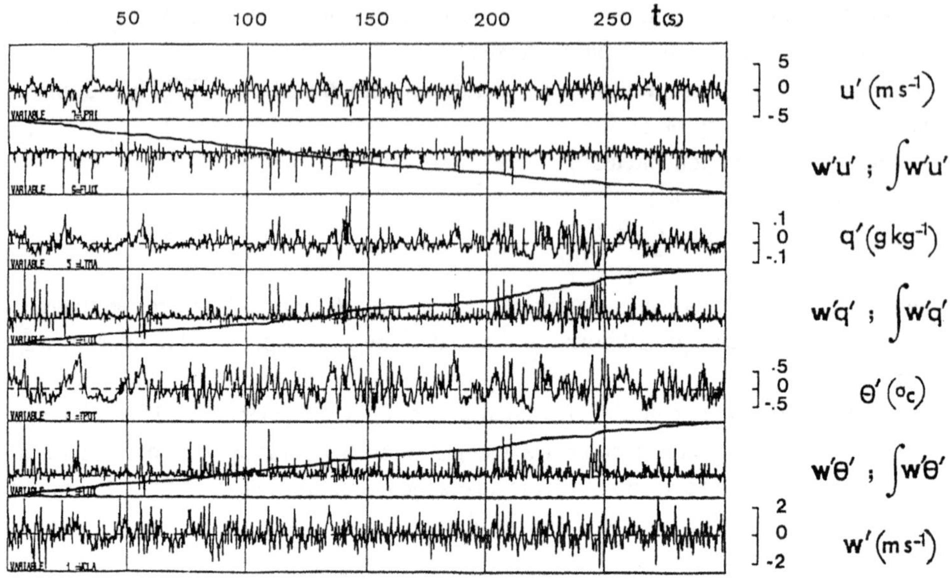

Fig. 3. Example of integration of turbulent fluxes: sensible heat ($w'\theta'$); latent heat ($w'q'$); and momentum ($w'U'$). Flight level: 30 m.

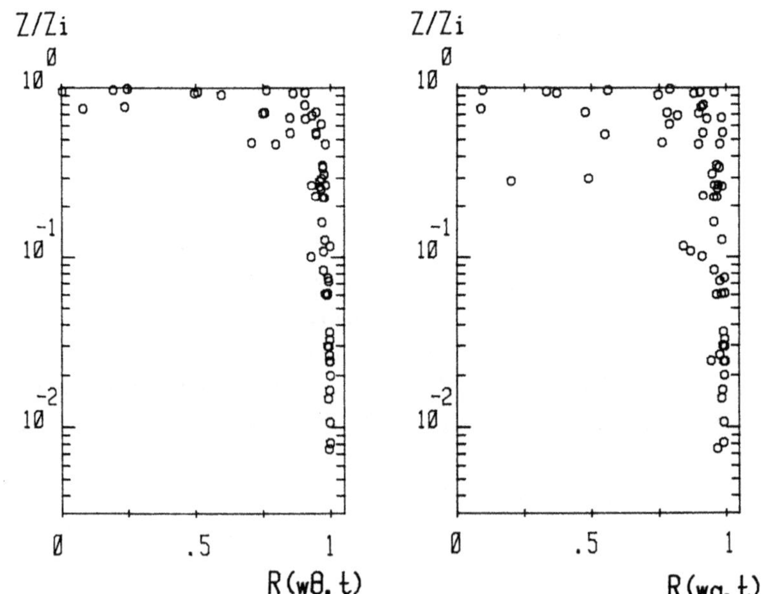

Fig. 4. Correlation coefficients between the time-integral of measured covariance and the straight line which best fits it, vs the reduced altitude z/Zi; sensible heat flux on the left and latent heat flux on the right sides of the figure.

4. Energy Budget

The data measured at the G.E.B. station allowed computation of the diurnal cycle of the energy budget. This budget will be compared with that of the Koorin experiment.

The airborne measurements, performed in the convective A.B.L. during the ECLATS experiment, will be compared with the two other budgets at midday.

4.1. The G.E.B. station results

The energy budget measurements were made over 25 days, using a sampling period of 20 min. The mean diurnal variation of the net radiative flux, the ground heat flux and the sensible heat flux are presented on Figure 5.

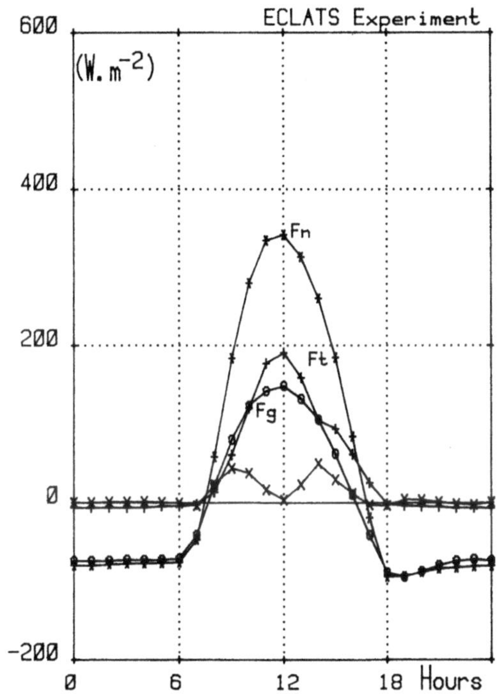

Fig. 5. Mean diurnal variation of the surface energy budget for the ECLATS experiment. Fn: radiative net flux; Ft: sensible heat flux; Fl: latent heat flux; Fg: ground heat flux.

4.1.1. *Net Radiative Flux (Fn)*

The maximum of the net radiative flux was 430 W m^{-2} and the mean maximum value was 340 W m^{-2}. The minimum value was observed at the beginning of the night. It could reach -114 W m^{-2} and on average -80 W m^{-2}. The average during the night was -68 W m^{-2}.

4.1.2. *Ground Heat Flux (Fg)*

The ground heat flux was an important term in the surface-layer energy budget. Its values were very high compared to those generally found in temperate regions where this term can often be neglected. The maximum was around midday, its mean value being

165 W m^{-2}. As generally expected for dry surfaces during the night, the net radiative flux was very close to the ground heat flux. (The latent heat flux was very small, and the sensible heat flux was of the order of -10 W m^{-2}).

4.1.3. *Sensible Heat Flux* (*Ft*)

The formulae used to compute the sensible heat flux depend on the stability of the A.S.L. ((6) and (7)). This stability can be characterized by the Richardson number Ri:

$$\mathrm{Ri} = \frac{g\, \partial\theta/\partial z}{T\,(\partial U/\partial z)^2} \,. \tag{12}$$

The hourly distribution of values of Ri encountered during the experiment, and the mean diurnal cycle, are presented in Figure 6. At night, values were between 0 and 0.2..

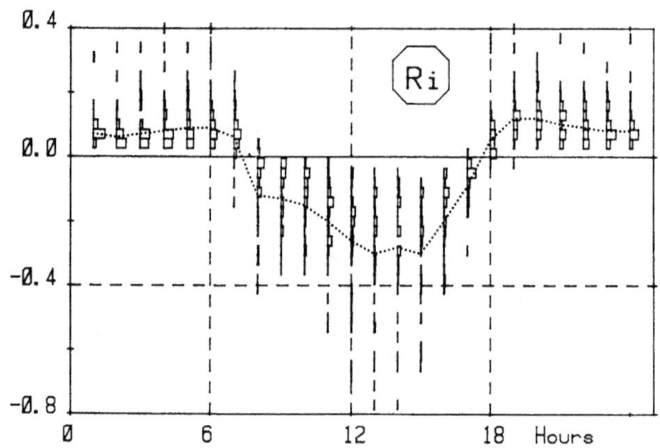

Fig. 6. Hourly distribution of Richardson gradient number over the whole period. Dotted line gives the best fit to the mean diurnal variation.

Values exceeding 0.2 were rare. They were associated with very weak windspeeds. The stability was greatest at the beginning of the night. In the morning, values were low, with small scatter (-0.2 on average), corresponding to high windspeeds (4 m s^{-1} at $z = 2.25$ m at 09 GMT). During the afternoon, the instability and its scatter increased (Ri = -0.35 on average), but Ri remained greater than -0.8. This period was characterized by lower windspeeds (3 m s^{-1} at 14 GMT). In the range of stability encountered ($-0.8 < \mathrm{Ri} < 0.2$), the universal functions used (ϕ_m and ϕ_T) were well defined.

The sensible heat flux was low during the night. Its mean value was -10 W m^{-2} but values sometimes dropped to -20 and -30 W m^{-2}. During the day, the sensible heat flux rose rapidly, lagging slightly behind the net radiative flux. Its variation in the afternoon showed assymmetry compared with that during the morning (and compared with the behavior of Fn and Fg), i.e., there was a rapid decrease in the flux at the beginning of the afternoon. The lowest values of Ri occurred at this time.

4.1.4. *Residual Term*

The residual term of the energy balance is the latent heat flux. In our case, values were small and the diurnal cycle was ill-defined, since estimates sum the errors in the other three components.

4.2. Comparison with the Koorin Experiment

Few data are available concerning the energy budget in tropical regions. However, the budget presented above can be compared with that obtained in the Koorin expedition (Clarke and Brook, 1979). This experiment took place in July and August 1974 at Daly Water in the north of Australia (16° 16′ S and 133° 23′ E), on bush land savanna. Garratt (1978a, b) describes the vegetation around the site of measurements:

- scattered live eucalyptus trees and shrubs (referred to as subtropical scrub savannah) of average height ~8 m (average crown diameter ~2 m) occupying some 25% of the total surface area;
- dry, dead grass of height <1 m, occupying some 60–70% of the surface area;
- patches of burnt grass and sandy soil making up no more than 10% of the area.

These surface conditions are very different from those of the G.E.B. station where the soil was bare.

The 30-day means of the ground energy budget have been described by Garratt (1978b). He also showed the effect of the canopy on the budget. We use here the ground

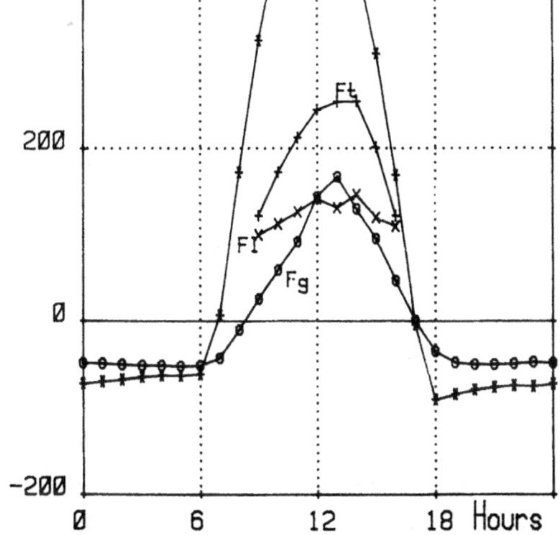

Fig. 7. Mean diurnal variation of the energy budget at ground level for the Koorin experiment. Fn: radiative net flux; Ft: sensible heat flux; Fl: latent heat flux; Fg: ground heat flux.

energy budget averaged over a period of 15 days (from day 14 to day 28) during which the cloud cover was slight.

The net radiative flux was measured at a height of 23 m, well above the trees, with a net radiometer (SRI). The sensible and latent heat fluxes were calculated at a height of 17.5 m by the eddy correlation method. The ground heat flux was measured using a heat flux plate (Middleton) (Clarke and Brook, 1979).

The mean budget is shown in Figure 7. The main differences with the ECLATS budget occur in the day-time. The net radiative flux, and the sensible and latent heat fluxes are greater at Koorin. During the night, the net radiative and the ground fluxes are of the same order of magnitude.

TABLE I

Comparisons of the various terms of the energy budget at midday for airborne measurements over the steppe (ECLATS), for the surface energy budget (G.E.B.) (ECLATS) and for Koorin. Fs: incoming short-wave radiation; Fn: radiative net flux; Ft: sensible heat flux; Fl: latent heat flux; Fg: ground heat flux; α: surface albedo; Rt: residual term. For Fn, the mean value is given at the measurement level (*Zm*) and the corrected ground-level value ($z = 0$ m).

	Airborne measurements ECLATS	G.E.B. station ECLATS	Koorin
Fs	803 ± 54	797 ± 38	842 ± 22
Fn	450 ± 35		538 ± 12
	30 m < Zm < 50 m		*Zm* = 23 m
Fn 0 m	424	340 ± 53	508
Ft	192 ± 35 (45%)	192 ± 30 (56%)	257 ± 46 (50%)
Fl	48 ± 20 (11%)	–	141 ± 62 (28%)
Fg	–	150 ± 29 (44%)	144 ± 12 (28%)
α	0.25 ± 0.02	0.36 ± 0.04	0.17
Rt	184 (Fg) (44%)	−2 (Fl) (0%)	−31 (−6%)

The main characteristics of the surface budget at midday are summarized in Table I for the two stations and as obtained from the airborne measurements. We have averaged the airborne measurements around 12 GMT at an altitude between 30 and 50 m in order to estimate the net radiative flux. In the same column, the sensible and latent heat fluxes are calculated as explained in Section 3.2.2. The residual term Rt is the non-computed term (for ECLATS) and includes the errors in the other terms. For airborne measurements, it represents the ground heat flux and for the G.E.B. station, the latent heat flux.

For both the Koorin experiment and the airborne ECLATS measurements, there was radiative flux divergence between the soil and the measurement level. For short-wave net flux, Cerf *et al.* (1982) showed that this attenuation was about 10 W m^{-2} for the highest concentration of dust haze encountered. Simulations of long-wave radiative flux profiles have been carried out by means of a radiative model described in Guedalia *et al.* (1984). The A.S.L. is described by a temperature profile deduced from the universal functions and according to the mean soil surface temperature and to the temperature profile of the A.B.L. For the ECLATS experiment, the mean soil surface temperature (Ts) deduced from an onboard radiometer (Barnes PRT5), assuming an emissivity of 1, was 43 °C (r.m.s. 3.1 °C); the mean air temperature above the A.S.L., converted adiabatically to ground level (Ta_0) was 31 °C (r.m.s. 2.2 °C). For the Koorin experiment, Ts was 44.2 °C (r.m.s. 1.3 °C) and Ta_0 was 26.8 °C (r.m.s. 0.9 °C). In both cases, the humidity profile was chosen independently of the altitude in the A.B.L. (4.5 g kg^{-1} for ECLATS and 7 g kg^{-1} for Koorin). The vertical profile of long-wave flux divergence is shown in Figure 8 for both situations. The attenuation occurs

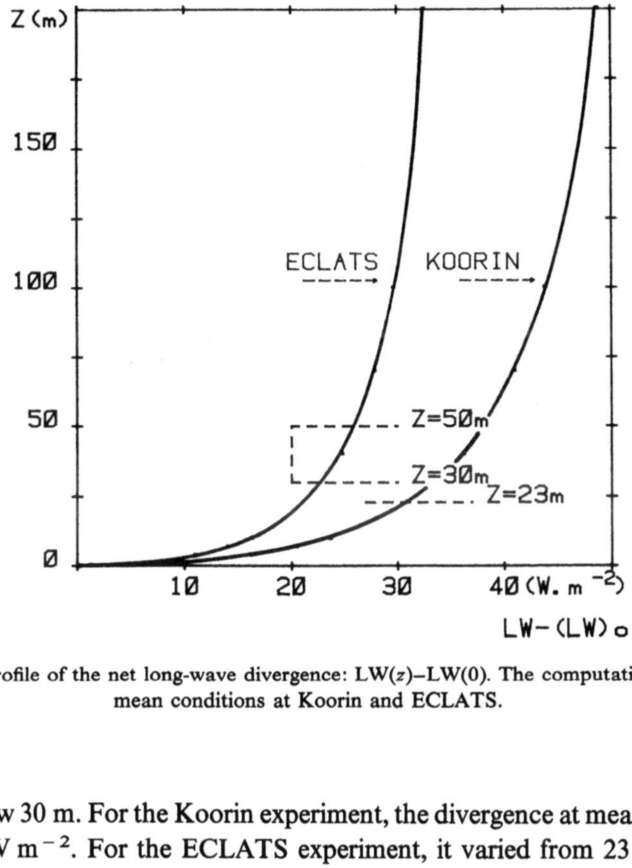

Fig. 8. Vertical profile of the net long-wave divergence: LW(z)–LW(0). The computation was made for mean conditions at Koorin and ECLATS.

principally below 30 m. For the Koorin experiment, the divergence at measurement level was about 30 W m^{-2}. For the ECLATS experiment, it varied from 23 to 27 W m^{-2} according to flight altitude; the correction assumed is 26 W m^{-2}. That leads to an adjustment of the net surface flux from 538 to 508 W m^{-2} for Koorin and from 450 to 424 for ECLATS.

In the three cases, the available energy at ground level (Fn) and its partition into heat fluxes (Ft, Fg, Fl) are very different. We first analyse the four components of the net radiative flux.

(1) The incoming short-wave radiation was about 40 W m^{-2} greater at Koorin than at ECLATS. In both experiments, the zenith angle was about the same. The difference is probably due to the high concentration of dust haze encountered during ECLATS (Fouquart *et al.*, 1985).

(2) The upward long-wave flux must have been about the same in the two experiments, because of the similar soil surface temperatures.

(3) The estimation of downward long-wave radiation was made using the model described above. The differences between the two experiments are due to the higher temperatures in the A.B.L. and to the dust haze in the ECLATS case. Two simulations were made, the first without dust, with the mean profiles of temperature and humidity of Koorin, the other with the mean profiles of temperature, humidity and dust of ECLATS. The comparison gives 40 W m^{-2} more in the case of ECLATS than of Koorin, at ground level. This estimation shows that this increase approximately balances the reduction of the short-wave incoming flux. Therefore, the differences in the net flux must be explained by the fourth term.

(4) The fourth term (upward short-wave radiative flux) is dependant on the surface albedo (α). For the flight area, the albedo was computed from the direct measurements of up- and down short-wave fluxes. The mean value is 0.25 (r.m.s. 0.02). At the G.E.B. station, local measurements (up and down long wave, downward solar and net radiative fluxes) lead to a mean value of 0.36 (r.m.s. 0.05). This difference can be explained by the areas seen by the sensors: the airborne measurements integrated a large area of steppe. At the G.E.B. station, the measurements were performed on bare soil, the surface characteristics of which were desert-like. In the case of the Koorin experiment, α was computed from measurements of downward short-wave and net fluxes. The net long-wave flux was estimated from the nocturnal radiative budget corrected by the diurnal variation of the soil temperature for upward long-wave flux and, for downward long-wave flux, by the radiative model. The estimated albedo is then 0.17. This computation is less accurate than the others, because we do not have direct measurements of long-wave radiative fluxes. This low value can be explained by the amount of evapotranspiration from vegetation in Koorin (140 W m^{-2} for latent heat flux).

The partition of the available energy (Fn) is made as follows: about half is converted into sensible heat flux towards the atmosphere (45% from airborne measurements, 56% from the G.E.B. measurements and 50% for Koorin). For ECLATS, the remaining half is used in ground heat flux (45% from airborne measurements and 44% from the G.E.B. measurements). For airborne measurements, 11% is used in latent heat flux. For Koorin, this other half is converted equally into ground heat flux and latent heat flux (28%). This partition is linked to the presence of active vegetation.

5. Conclusion

A set of data has been presented concerning the energy budget of the Sahel A.S.L. These data were collected during the ECLATS experiment from local energy budget measurements (G.E.B. station) and from low-level aircraft measurements performed over the steppe. The latter demonstrate the spatial homogeneity of the fluxes over this region.

The energy budget has been compared with the one established during the Koorin expedition (local measurements). The diurnal cycles show similar behavior in the night-time. During the day, the mean budgets at midday show the following differences:

the incoming short-wave flux is greater (about 40 W m^{-2}) at Koorin than in ECLATS, mainly because of the greater optical thickness of the atmosphere (dust haze in ECLATS). This is counterbalanced by the increase of downward long-wave flux in ECLATS due to the dust haze and to the higher air temperature in the A.B.L. Therefore, differences in the net flux are due to the surface albedo α, which is 0.36 at the G.E.B. station. This value is characteristic of the bare soil in the vicinity of the measurement point. As for airborne measurements, the mean value is 0.25, characterizing the Sahel bushy steppe in the dry period. For the Koorin experiment, we found a smaller albedo (0.17), because of the transpiring vegetation.

Half of the available energy at midday (424 W m^{-2} for the Sahel steppe, 340 W m^{-2} at the G.E.B. station and 510 W m^{-2} for Koorin) is used for sensible heat flux towards the atmosphere. The other half, for ECLATS, is used principally for ground heat flux; for Koorin, this half is equally divided between the latent heat flux and the ground heat flux. In the Sahel region, the evaporation is very weak (close to 0 at the G.E.B. station and 11% over the steppe).

Acknowledgements

The ECLATS Experiment was financially supported by the 'Action Thématique Programmée: Recherches Atmosphériques' of the 'Institut National d'Astronomie et de Géophysique' (INAG) and by the 'Direction des Recherches: Etudes et Techniques', of the Defence Ministry. We express our gratitude to the Niger authority which authorized this expedition. We would like to thank people from the universities of Niamey (Niger), Abidjan (Cote d'Ivoire), Lille and Toulouse (France). Particular thanks to the 'Division Technique' of the INAG, for their experimental support, to D. Guedalia and C. Estournel for their contribution to the analysis of radiative effects, to B. Benech for his comments and to A. Herrada for her contribution to the numerical analysis.

References

Bertrand, J., Baudet, J., and Drochon, A.: 1975, 'Importance des aérosols naturels en Afrique de l'Ouest', *J. Rech. Atmos.* **8**, 845–850.

Cerf, A., Fouquart, Y., Bonnel, B., Broniez, B., Chaoui-Roquai, M., and Smith, L.: 1982, 'Propriétés radiatives de l'aérosol sahélien, *La Météorologie VI* **29–30**, 247–260.

Clarke, R. H. and Brook, R. R.: 1979, *The Koorin Expedition: Atmospheric Boundary-Layer Data over Tropical Savannah Land*, Department to Science and The Environment, Bureau of Meteorology Australian Government Publishing Service, Canberra.

Druilhet, A. and Tinga, A.: 1982, 'Présentation de l'expérience ECLATS', *La Météorologie VI* **29–30**, 203–212.

Druilhet, A. and Durand, P.: 1984, 'Etude de la couche limite convective sahélienne en présence de brumes sèches (expérience ECLATS)', *Boundary-Layer Meteorol.* **28**, 51–77.

Durand, P.: 1983, 'Etude des caractéristiques turbulentes de la couche limite convective sahélienne (Expérience ECLATS)', Thèse de Docteur-Ingenieur, Univ. de Toulouse III, n° 826, 165 p.

Dyer, A. J. and Hicks, B. B.: 1970, 'Flux-Gradient Relationships in the Constant Flux Layer', *Quart. J. Roy. Meteorol. Soc.* **96**, 715–721.

Estournel, C., Vehil, R., and Guedalia, D.: 1986, 'An Observational Study of Radiative and Turbulent Cooling in the Nocturnal Boundary Layer (ECLATS Experiment)', *Boundary-Layer Meteorol.* **34**, 55–62.

Fouquart, Y., Cerf, A., Bonnel, B., Brogniez, G., Buriez, J. C., and Smith, L.: 1986, 'Observation of Saharan Aerosols: Results, of ECLATS Field Experiment, Part II: Optical Properties of Aerosols', *J. Clim. Appl. Meteorol.* (to appear).

Garratt, J. R.: 1978a, 'Flux Profile Relations Above Tall Vegetation', *Quart. J. Roy. Meteorol. Soc.* **104**, 199–211.

Garratt, J. R.: 1978b, 'Transfer Characteristics for a Heterogeneous Surface of Large Aerodynamic Roughness', *Quart. J. Roy. Meteorol. Soc.* **104**, 491–502.

Guedalia, D., Estournel, C. and Vehil, R.: 1984, 'Effects of Sahel Dust Layers Upon Nocturnal Cooling of the Atmosphere (ECLATS Experiment)', *J. Clim. Appl. Meteorol.* **23**, 644–650.

Oladipo, E. O.: 1980, 'An Analysis of Heat and Water Balances in West Africa', *Geograph. Rev.* **70**, 194–209.

Paulson, C. A.: 1970, 'The Mathematical Representation of Wind Speed and Temperature Profiles in the Unstable Atmospheric Surface Layer', *J. Appl. Meteorol.* **9**, 857–861.

Riou, C., Lagouarde, J. P., and Chartier, R.: 1979, 'Evaporation du sol nu en zone semi-aride et en conditions hivernales. Relations avec l'albedo et la temperature de la surface du sol', *Ann. Agron.* **30**, 47–361.

Webb, E. K.: 1970, 'Profile Relationships: The Log-Linear Range and Extension to Strong Stability', *Quart. J. Roy. Meteorol. Soc.* **96**, 67–90.

THE PERSISTENCE OF POLLUTANTS DOWNWIND OF A POINT SOURCE FOLLOWING TERMINATION OF THE EMISSION

ROLAND R. DRAXLER

Air Resources Laboratory, National Oceanic and Atmospheric Administration, Silver Spring, MD, U.S.A.

(Received in final form 2 January, 1987)

Abstract. Experimental studies have shown that tracers injected over short time intervals can often be measured downwind near the ground for a much longer time than would be expected considering the duration of the release and the wind speed. A two-dimensional advection-diffusion model was used in a numerical study to investigate this effect. The results from the numerical study showed that the rate of decrease in concentrations downwind at a receptor after a source has stopped emitting was more sensitive to magnitude of the wind speed and alongwind mixing coefficient than the rate of increase in concentrations when the source first starts emitting. The lower the wind speed or the greater the mixing, the more slowly the concentration decreased as the region upwind of the receptor continued to contribute tracer until it was depleted by advection and vertical diffusion. These atmospheric processes are able, qualitatively, to account for the slowly decreasing air concentrations.

1. Introduction

The crosswind and vertical dispersion of pollutants from continuous sources is probably one of the most investigated topics in air quality modeling. Various approaches used to define the continuous plume concentration distributions have been summarized by Pasquill and Smith (1983) and more recent developments have been addressed by Weil (1985). As many of the atmospheric dispersion experiments conducted in recent years were designed to measure only crosswind and vertical dispersion, relatively few data exist (Draxler, 1979, 1984) describing alongwind dispersion. This is in part due to the greater interest in continuous sources in which alongwind dispersion may be neglected. However, when modeling quasi-instantaneous sources, or 'puffs', the horizontal growth in both crosswind and alongwind directions becomes equally important.

The vertical shear of the horizontal wind gives a much more complex structure to a pollutant 'puff' than the circular distributions (alongwind dispersion equals crosswind dispersion) used in many models. Numerical simulations by Taylor (1982) showed that at longer travel times (24 hr), the alongwind spread can exceed the crosswind spread by as much as a factor of five due to the effects of wind shear. At shorter times (3 hr), the alongwind spread only exceeded the crosswind spread by a factor of two.

A more typical emission would probably be somewhere between an instantaneous and a continuous one, e.g., that from an industial process where emissions occur only during working hours. In that case, the pollutant transport and dispersion might be simulated by a 'segmented' plume model (Lott, 1984), where the continuous source plume has a discrete beginning and end. The degree of alongwind dispersion can have a considerable effect upon the air concentrations downwind of such a source. The data from a tracer experiment conducted for over one year, where a perfluorocarbon tracer source was sequentially on for six hours and then off for 30 hr, were used to examine the time history of air concentrations downwind of the source. Draxler (1987) found that 8-hr averaged tracer air concentrations, at samplers which were on average 24 km downwind, tended to remain high well after the tracer release terminated. The air concentrations decreased at a rate of 20% per hour. The long persistence of the tracer was attributed to the combined effect of little vertical mixing and slower advection winds near the surface, where high tracer concentrations had accumulated during the ground-level release. In this paper, a two-dimensional advection-diffusion model will be used to show how the persistence of pollutants from ground-level sources can be explained by the wind and vertical mixing profiles as well as the alongwind dispersion.

2. Advection-Diffusion Model

The pollutant is assumed to be emitted near the ground. The vertical profiles of the horizontal wind (U) and vertical mixing coefficient (K) are known and variations in density are neglected. As the alongwind mixing coefficient (H) is not well understood, it is specified as a constant with distance and height. This entire process is simulated by the two-dimensional advection-diffusion equation, where the local rate of change of the pollutant mass is given by

$$\frac{\partial Q}{\partial t} = -\mathbf{V}\cdot\nabla Q + \frac{\partial}{\partial Z}\left(K\,\frac{\partial Q}{\partial Z}\right) + \frac{\partial}{\partial X}\left(H\,\frac{\partial Q}{\partial X}\right). \tag{1}$$

Equation (1) is solved on a rectangular grid with i and k coordinates increasing downwind (X) and upward (Z), respectively. Horizontal gradients are computed from upwind differences and vertical gradients are computed from centered differences. The vertical velocity is assumed to be zero so that the vector wind $\mathbf{V}$ reduces to just a horizontal wind U; then the computational formula for Equation 1 becomes:

$$\begin{aligned} Q_{i,k}(t+\Delta t) = Q_{i,k}(t) - \Delta t[&U_k(Q_{i,k} - Q_{i-1,k})/\Delta X + \\ &+ \frac{K_{\hat{k}+1}(Q_{i,k+1} - Q_{i,k})}{(Z_{k+1} - Z_k)(Z_{\hat{k}+1} - Z_{\hat{k}-1})} - \frac{K_{\hat{k}-1}(Q_{i,k} - Q_{i,k-1})}{(Z_k - Z_{k-1})(Z_{\hat{k}+1} - Z_{\hat{k}-1})} + \\ &+ H(Q_{i+1,k} - Q_{i,k})/\Delta X^2 - H(Q_{i,k} - Q_{i-1,k})/\Delta X^2]\,. \end{aligned} \tag{2}$$

In Equation (2) the horizontal wind speed and Q are defined at the central height of each grid box. Winds are the same at each k for all i. The vertical mixing coefficient is defined at the height of the interface between adjacent boxes, and the upper and lower boundaries of box k are denoted by $\hat{k}+1$ and $\hat{k}-1$, respectively. The vertical indices are defined such that $k = \hat{k}$ and thus the upper interface of box k, $\hat{k}+1$, equals the lower interface of box $k+1$. The horizontal grid distance, ΔX is the same for all i; however, the vertical grid spacing increases with height as shown in Figure 1 by the horizontal dashed lines at each box interface.

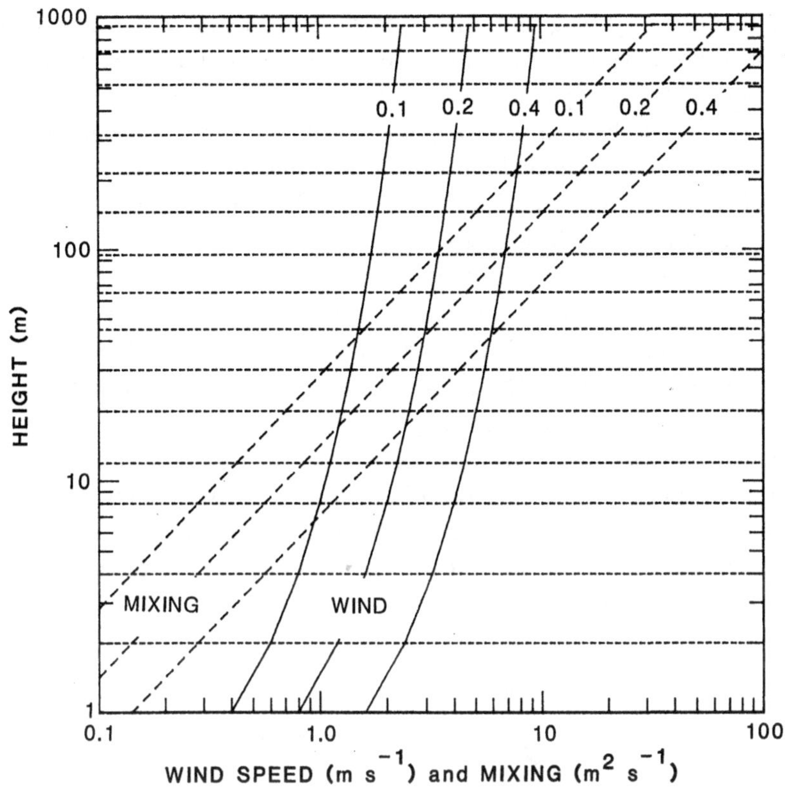

Fig. 1. Profiles of horizontal wind and vertical mixing used for the three simulation cases of $u_* = 0.1, 0.2$, and 0.4 m s^{-1}. Vertical grid spacing is shown as the small dashed horizontal lines.

At the upwind boundary ($i = 1$), $Q = 0$ at all times. The mass, Q, at the pollutant source box is increased by the product of the emission rate (q) and time step (Δt) at each iteration before solving Equation (2). Full reflection is assumed at the upper and lower boundaries of the model so that K at those interfaces equals zero.

The wind and mixing profiles are defined by those occurring during neutral stability conditions such that

$$U = \frac{u_*}{0.35} \ln\left(\frac{z}{z_0}\right) \tag{3}$$

and

$$K = 0.35 u_* z , \tag{4}$$

where u_* is the friction velocity and z_0 is the roughness length. Model simulations were performed for three different friction velocities (0.1, 0.2, 0.4 m s^{-1}) all with $z_0 = 0.25$ m. The profiles of U and K for the three values of u_* are shown in Figure 1. An approximate logarithmic vertical spacing was chosen to enable larger time steps to be used in the integration. Time steps (Δt) of 20, 12, and 6 s were required to maintain computational stability for each increase in friction velocity.

The vertical grid spacing and mixing coefficient determine the time step but the horizontal grid spacing influences the undesirable effects of artificial diffusion. As the main purpose of this study is to examine the rate of concentration change during the transition periods of the source turning on and off, the effect of the artificial diffusion on the advection term is especially important. The numeric calculations were performed for four different ΔX values: 0.25, 0.5, 1.0, and 2.0 km, using a value of 0.1 m s^{-1} for u_*. Since the horizontal diffusion term will mask the effects of artificial diffusion, H was set to zero for these trials. Values of air concentration normalized by the emission rate are shown in Figure 2 as a function of time for the four grid sizes at a distance of 24 km from the source for the lowest grid box. The vertical dashed line on Figure 2 shows when the emission was terminated. The source is started at $t = 0$ and maintained by adding $q\Delta t$ into the source grid box until a steady-state is achieved at about $t = 12$ hr. Although Q no longer increases significantly after about six hours, a steady-state is defined to exist when the advection and diffusion terms of Equation 2 cancel each other to at least three significant figures. At this point, the emissions are stopped and soon thereafter the concentration begins to decrease. Clearly shown on Figure 2 is the much larger effect of grid size on the side of decreasing concentration rather than the region of initial

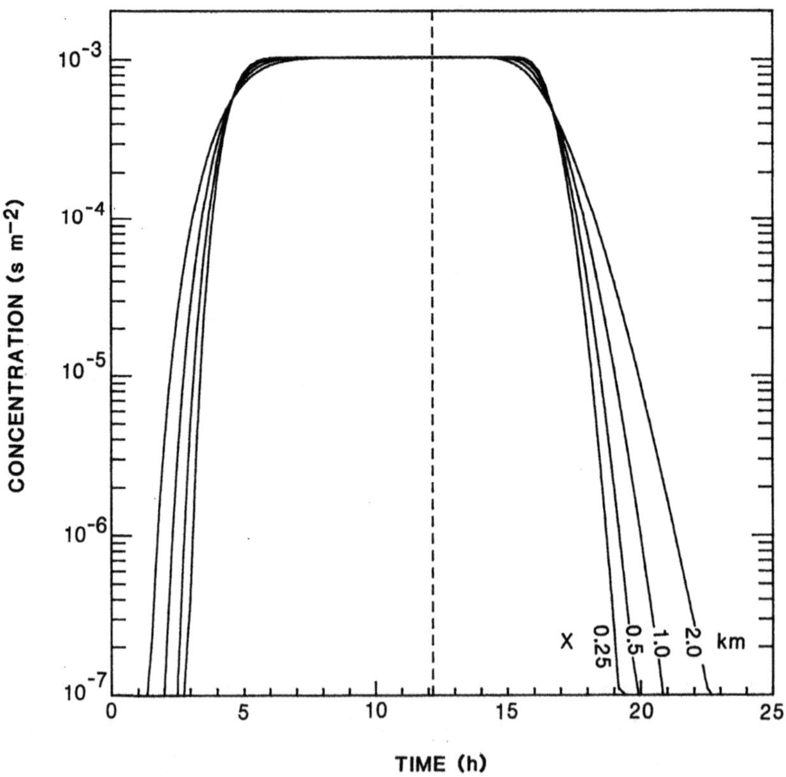

Fig. 2. Time history of air concentration in the lowest grid box 24 km downwind for four values of horizontal grid spacing (0.25, 0.5, 1.0, and 2 km). Dashed vertical line indicates when the source has been turned off.

concentration rise. As smaller values of ΔX require a larger grid and greater computational time, subsequent calculations will be performed with a horizontal grid spacing of 0.5 km. This results in a reasonable 15 (900 m) by 50 (25 km) calculational grid while keeping the effects of artificial diffusion small.

Another check on the numeric calculation is to compare the steady-state results with the analytic solution for the crosswind integrated concentration (CWIC). This would be the same result as when Q from the two-dimensional model is divided by the $\Delta Z \Delta X$ of the grid box to obtain concentration. The most simple case is to assume a constant profile of U and K; then the ground-level CWIC,

$$C(X) = qU^{-1}(\pi KX/U)^{-0.5}, \tag{5}$$

where X is downwind distance and as before, q is the emission rate. For the most realistic comparison to the profiles of U and K, constant values of 1.0 m s^{-1} and 1.0 m^2 s^{-1}, respectively, are used; these are about equal to the median values for profiles of $u_* = 0.1$ m s^{-1} (see Figure 1). The simulation with U and K constant with height and $H = 0$ achieved a steady state after 700 min. The ratio of the concentration from Equation (5) to the numeric solution was better than 1.01 at 24 km downwind. Of course at the first grid box downwind of the source, the numeric solution was low by a factor of two. The two solutions quickly converged within several grid points. From the previous results, one can presume that the numeric technique, selection of grid spacing, and time step does not introduce any significant bias when the results are restricted to the farther downwind grid boxes.

3. Results

The motivation for this study is to evaluate changes in ground-level air concentrations downwind of a continuous source after the source starts and stops. In regard to the discussions in the introduction, this is similar to the passage of a quasi-instantaneous puff over a receptor. Experimental data (Draxler, 1987) indicated that concentrations decreased exponentially downwind of the source after the emissions stopped. The exponential decrease as well as an exponential rise is evident for the numeric calculations shown in Figure 2. All subsequent calculations are shown for a receptor distance of 24 km, a distance chosen to be representative of the average source receptor distance in the field experiment that initiated this study.

The time history of concentration at a receptor is examined by performing the numeric calculations with various wind and mixing profiles. In the first group of simulations, to test the effect of wind speed, the mixing coefficient profile was held constant ($u_* = 0.1$ m s^{-1}) for the three different wind speed profiles ($u_* = 0.1$, 0.2, and 0.4 m s^{-1}). In the following simulation, the horizontal mixing coefficient was set to zero, although there will still be some numerical diffusion. The results shown in Figure 3, as well as in all other figures, are for the lowest grid box (2 m) 24 km downwind from the source box. In the simulation shown on Figure 3, the emissions were stopped at about 9 hr, shown as the vertical dashed line.

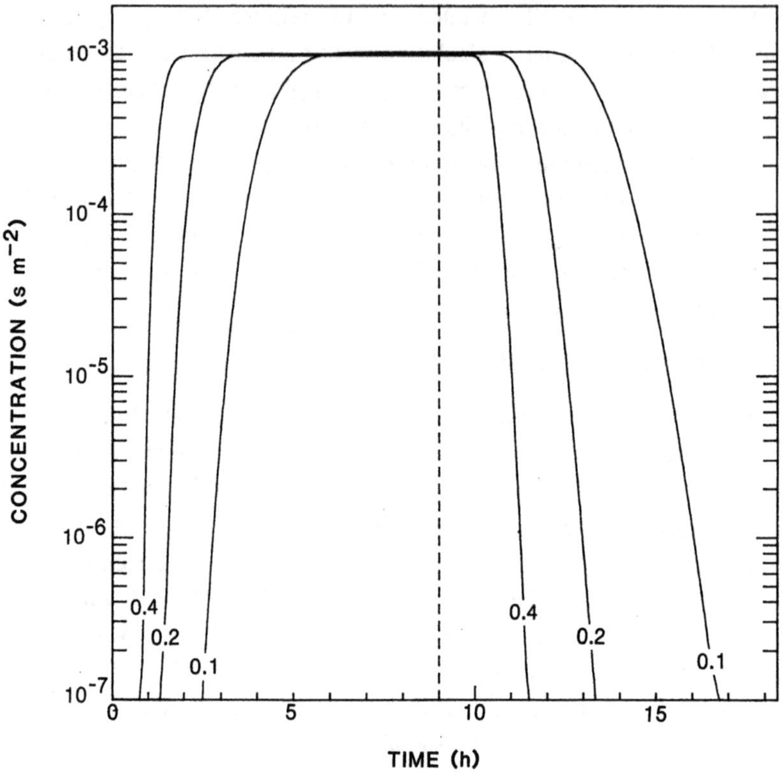

Fig. 3. Time history of air concentration in the lowest grid box 24 km downwind for the three values of wind speed ($u_* = 0.1$, 0.2, and 0.4 m s^{-1}) with the mixing held constant at $u_* = 0.1$ m s^{-1}. The source started 0 hr and stopped at 9 hr as indicated by the vertical dashed line.

One of the more striking results is the apparent independence of the steady-state concentration, from changes in wind speed. Although wind speeds double with each increase in friction velocity, the normalized steady-state concentration (about 10^{-3} s m^{-2}) only decreases 3% with each increment of u_*. When the simulations were run with different wind speeds but with U and K constant with height, a result similar to Equation (5) was obtained, where the steady-state concentration was inversely proportional to wind speed, a well-known property of the Gaussian plume model. Evidently the fact that U and K increase with height alters the model sensitivity to these parameters. Similar results were obtained by Sutton (1953, p. 281) using a power law profile for U and K typical of neutral conditions. His analytic solution indicated that the concentration was inversely proportional to only the 1/9 power of wind speed, which implies about an 8% change in concentration for each doubling of wind speed.

The other feature of Figure 3 is the similar rate of concentration increase and decrease. For each wind speed profile, the rate of concentration increase and decrease is about the same while the slopes are progressively steeper with greater wind speeds. Note that the rapid decrease in concentration lags more, with each drop in wind speed, than the initial concentration rise; the difference in lag is more pronounced at lower wind speeds. This is due in part to the greater advection time with each lower friction velocity. In fact, the concentration did not return to its initial value (10^{-7} s m^{-2}) until a time after release termination that was about equal to X/U in the lower grid boxes. However, the travel time to the initial rise was about equal to X/U in the upper model regions. This suggests that air quality models designed to calculate temporal variations of ground-level air concentrations need to incorporate the wind and mixing profiles rather than to use one wind to represent the vertical layer through which the pollutant has mixed.

The difference between the rapid rise of concentration and slower decrease was examined further by separating the advection and diffusion terms in Equation (2). The vertical and horizontal mass flux is shown in Figure 4 for various times corresponding with the same u_* curves as used in Figure 3. The curves show the percent change in mass as a function of time in the lowest grid box 24 km downwind due to the vertical diffusion term (solid line) and advection term (dashed line). The curves are marked with each friction velocity. The solid and dashed curves all have corresponding peaks at the same time. For instance, if we examine the third peak shown in Figure 4, those for $u_* = 0.1$ m s^{-1}, the initial rise in concentration is dominated by contributions from vertical diffusion, whose peak value is twice the advection term. The fact that the vertical diffusion term dominates the local rate of change of concentration confirms the previous assumption that the initial downwind concentration increase is dominated by the

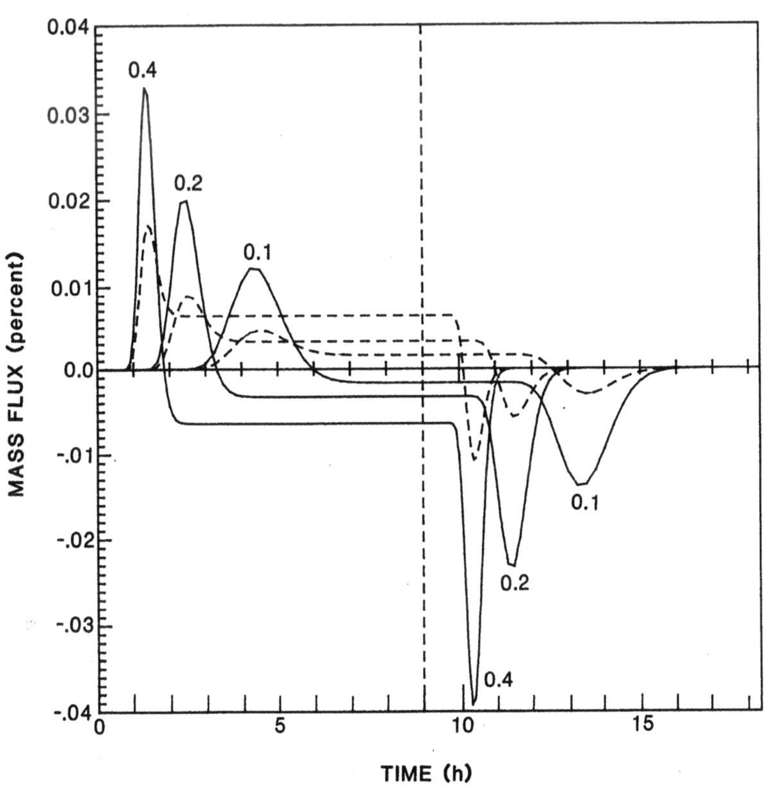

Fig. 4. Time history of the advection (dashed) and diffusion (solid) terms of Equation (2). The three groups of curves are for the three friction velocities as shown in Figure 3.

transport at heights well above the ground; the diffusion contribution to the lowest box can only come from aloft.

After the source has stopped at 9 hr, steady-state conditions persist for another 3 hr; that is, advection contributes mass while diffusion removes mass. The two processes of course cancel each other, the definition of a steady-state. This condition persists after the source has stopped for a period equal to the fastest advection time (3 hr), from the upper-level wind speeds. At this point, the advection term becomes negative and both advection and diffusion losses increase to a maximum and then rapidly return toward zero as all the mass is eventually lost. During the initial rise of concentration, the vertical concentration profile increases with height, so that mass can mix downward to achieve eventually a steady-state profile which decreases with height. The three sets of curves show how vertical mixing dominates those periods with rapid temporal changes in concentration. During the steady-state period, however, the balance between advection and vertical mixing to the ground determines the concentration.

As vertical diffusion dominates the period of most rapid change at a receptor, the calculations were performed with various vertical mixing profiles while holding the wind speed profile constant at $u_* = 0.1 \text{ m s}^{-1}$. The time history of concentration is shown on Figure 5 for these simulations. Here doubling the mixing ($u_* = 0.1$ to 0.2) halves the steady-state concentration. A further doubling of mixing (0.2 to 0.4) has little effect. This

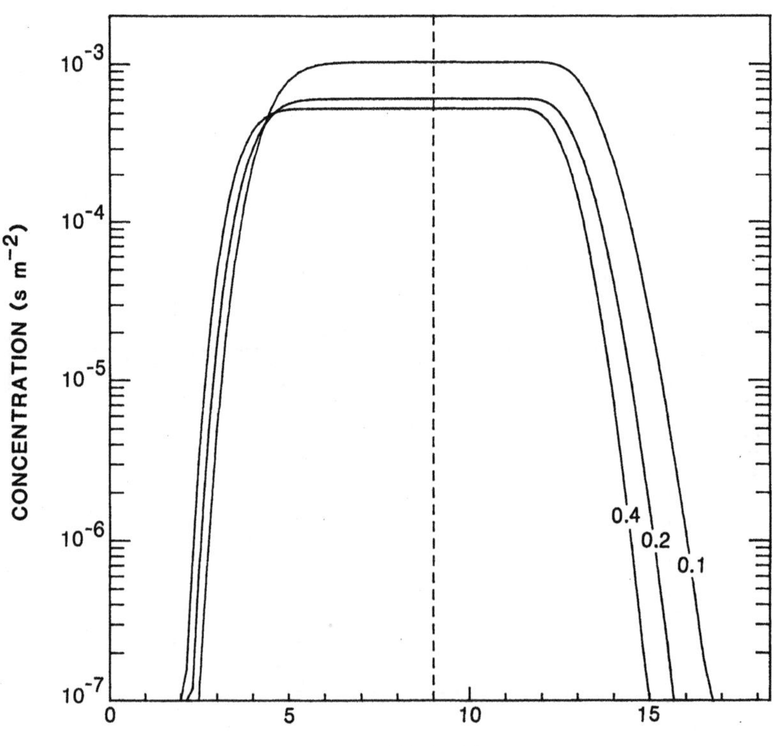

Fig. 5. Time history of air concentration as in Figure 3 with various mixing profiles ($u_* = 0.1$, 0.2, and 0.4 m s^{-1}) and with wind speed held constant at $u_* = 0.1 \text{ m s}^{-1}$.

is due to the full reflection at 900 m at the upper model boundary. After some critical value, further mixing will not affect a well-mixed vertical distribution. However, in terms of the rate of concentration rise and fall, doubling the vertical mixing is not as significant as doubling wind speed; hence, it is possible to conclude that wind speed differences dominate the temporal history of a pollutant over a receptor while the vertical mixing coefficient is more important in determining the magnitude of the steady-state concentration.

The last area of concern is the effect of horizontal mixing. The previous results were all calculated with no horizontal mixing. Simulations with a constant horizontal mixing coefficient of 0, 10^3, 2×10^3, 5×10^3, and 10^4 m^2 s^{-1} are shown in Figure 6 with profiles of U and K for $u_* = 0.1$ m s^{-1}. Except for the fact that the steady-state concentration decreases with increased horizontal mixing, the curves are very similar to those shown in Figure 2, where the effect of various grid sizes was examined. The drastic change in slope of the concentration decrease region shows how these concentrations are most sensitive to variations in horizontal mixing as well as wind speed. The high concentration values persist longer and do not decrease as quickly because of the continued advection and mixing of material that has diffused upwind.

The theoretical calculation can qualitatively be compared with actual measurements. In the year-long experiment cited earlier in which a tracer was released at regular

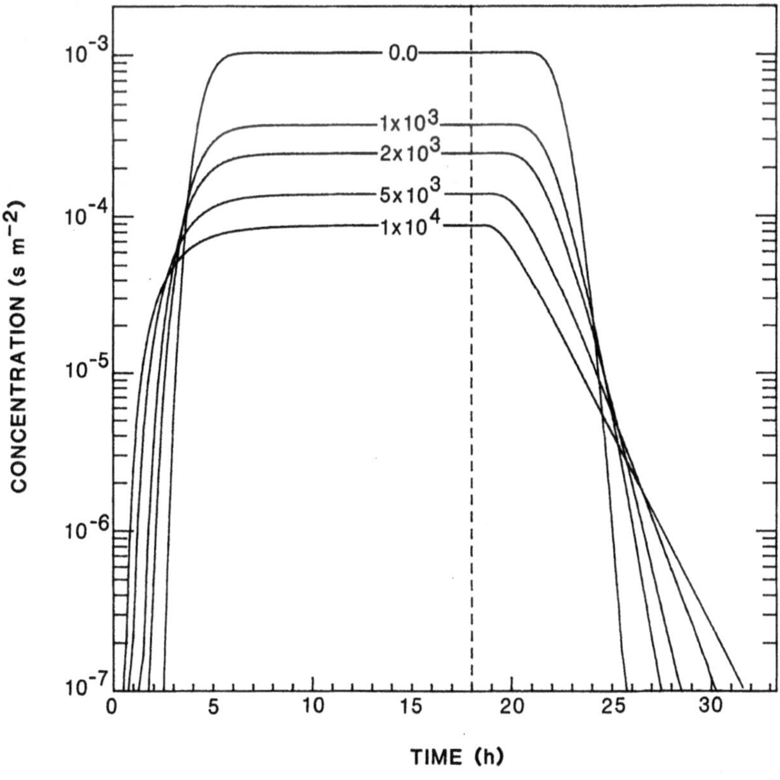

Fig. 6. Time history of air concentration where U and K profiles are for $u_* = 0.1$ m s^{-1} and the horizontal diffusion coefficient was varied between 0 and 10^4 m^2 s^{-1}.

intervals (36 hr) and measured continuously at 8 hr intervals, Draxler (1987) found many cases of air concentration remaining high many hours after the tracer release stopped. The concentration measurements decreased at about 20% per hour. The data shown on Figure 6 are somewhat representative of actual conditions. For the curve with the greatest horizontal mixing, the model concentration decreases twice as fast (40% per hour) as the concentrations did during the experiment. The model calculation is only a crude approximation to the more complicated processes occurring in the atmosphere. If in the model calculation, with a flat surface without obstructions, mass could persist for such long periods of time (14 hr) near the surface, it is all the more likely that in the atmosphere with many obstructions, such as buildings and vegetation, real pollutants could be trapped in various stagnation pockets and be seen downwind for extended periods while these pockets are slowly depleted.

4. Summary

A two-dimensional advection-diffusion model was used to simulate the behaviour of a pollutant downwind of a continuous source when that source stops emitting. This question arises from the analysis of experimental data where perfluorocarbon tracers were used to simulate real sources. As the tracer persisted over the samplers for longer times than expected, questions were raised regarding whether the persistence was due to properties of the tracer or the nature of atmospheric diffusion. In general, the numeric simulation showed that the pollutant concentration decreased with time more slowly than it increased when the source started emitting although the rate of decrease was about the same as the increase when there was no horizontal mixing. The time it took the pollutant to reach background levels after the source stopped emitting was much more sensitive to wind speed than the time it took for the pollutant to appear at the receptor when the source started. This is due to the fact that the first appearance was controlled by the upper-level winds while the lag of concentration over the receptor was controlled by how quickly the upwind mass was depleted by slower low-level winds. A similar result was achieved, a much slower decrease in concentration at a receptor, as the horizontal mixing coefficient increased. Increased horizontal mixing resulted in more pollutant diffusing upwind to be later advected over the sampler. Simulations of the steady-state concentration showed that it was only sensitive to the vertical mixing coefficient and not wind speed. This occurred only when the mixing coefficient increased with height. A qualitative comparison of the model calculation to some experimental data showed similar persistence of concentration downwind of the former source. The results would suggest that air quality models should not assume symmetric upwind-downwind pollutant distributions, but they should be skewed in the upwind direction to account for the material lagging in the lower layers of the atmosphere where winds and mixing are much less. The persistence of concentration over the receptor was about equal to the travel time from source to receptor in the lowest layer near the ground.

References

Draxler, R. R.: 1979, 'Some Observations of the Along-Wind Dispersion Parameter', Preprint, 4th Symp. on Turb., Diffusion, and Air Pollution, January 15–18, Reno, NV, American Meteorological Society, Boston, pp. 5–8.

Draxler, R. R.: 1984, 'Diffusion and Transport Experiments', Chapter 9, in D. Randerson (ed.), *Atmospheric Science and Power Production*, DOE/TIC-27601, pp. 367–422.

Draxler, R. R.: 1987, 'One Year of Tracer Dispersion Measurements over Washington, D.C.', *Atm. Environ.* **21**, 69–77.

Lott, R. A.: 1984, 'Performance of a Time-Dependent Dispersion Model in Elevated Terrain', *Atm. Environ.* **18**, 2367–2376.

Pasquill, F. and Smith, F. B.: 1983, *Atmospheric Diffusion*, 3rd ed., Ellis Horwood, Chichester, U.K., 347 p.

Sutton, O. G.: 1953, *Micrometeorology*, McGraw-Hill, Book Co., Inc., New York, 333 p.

Taylor, A. D.: 1982, 'Puff Growth in an Eckman Layer', *J. Atm. Sci.* **39**, 837–850.

Weil, J. C.: 1985, 'Updating Applied Diffusion Models', *J. Climate Appl. Meteorol.* **24**, 1111–1130.

NON-DIMENSIONAL WIND AND TEMPERATURE PROFILES IN THE ATMOSPHERIC SURFACE LAYER: A RE-EVALUATION

ULF HÖGSTRÖM

Department of Meteorology, Box 516, S-751 20 Uppsala, Sweden

(Received in final form 27 February, 1987)

Abstract. Previous results of non-dimensional wind and temperature profiles as functions of $\zeta(=z/L)$ show systematic deviations between different experiments. These discrepancies are generally believed not to reflect real differences but rather instrumental shortcomings. In particular, it is clear that flow distortion has not been adequately treated in most previous experiments. In the present paper, results are presented from a surface-layer field experiment where great care was taken to remove any effects from this kind of error and also to minimize other measuring errors. Data from about 90 30-min runs with turbulence measurements at three levels (3, 6, and 14 m) and simultaneous profile data have been analysed to yield information on flux-gradient relationships for wind and temperature.

The flux measurements themselves show that the fluxes of momentum and sensible heat are constant within $\pm 7\%$ on average for the entire 14 m layer in daytime conditions and when the stratification is slightly stable. For more stable conditions, the flux starts to decrease systematically somewhere in the layer 6 to 14 m. From a large body of data for near-neutral conditions ($|\zeta| \leq 0.1$), values are derived for von Kármán's constant: 0.40 ± 0.01 and for ϕ_h at neutrally, 0.95 ± 0.04. The range of uncertainty indicated here is meant to include statistical uncertainty as well as the effect of possible systematic errors.

Data for ϕ_m and ϕ_h for an extended stability range ($1 > \zeta > -3$) are presented. Several formulas for ϕ_m and ϕ_h appearing in the literature have been used in a comparative study. But first all the formulas have been modified in accordance with the following assumptions: $k = 0.40$ and $(\phi_h)_{\zeta=0} = 0.95$; deviations from this result in the various studies are due to incomplete correction for flow distortion. After new corrections are introduced, the various formulas were compared with the present measurements and with each other. It is found that after this modification, the most generally used formulas for ϕ_m and ϕ_h for unstable conditions, i.e., those of Businger *et al.* (1971) and Dyer (1974) agree with each other to within $\pm 10\%$ and with the present data. For stable conditions, the various formulas still disagree to some extent. The conclusion in relation to the present data is not as clear as for the unstable runs, because of increased scatter. It is, however, found that the modified curve of Businger *et al.* (1971) for ϕ_h fits the data well, whereas for ϕ_m, Dyer's (1974) curve appears to give slightly better agreement.

1. Introduction

It has long been accepted within the scientific community that wind and temperature profiles in the horizontally homogeneous surface layer properly normalized with scales for velocity and temperature should be unique functions of z/L, i.e.:

$$\frac{\partial \bar{u}}{\partial z} \frac{kz}{u_*} = \phi_m(z/L), \tag{1}$$

$$\frac{\partial \bar{\theta}}{\partial z} \frac{kz}{T_*} = \phi_h(z/L); \tag{2}$$

Boundary-Layer Meteorology **42** (1988) 55–78.

with the following notations:

$\bar{u}$	magnitude of the horizontal wind vector,
$\bar{\theta}$	potential temperature,
z	vertical coordinate,
u_*	friction velocity = τ/ρ_0,
T_*	scaling temperature = $-\overline{w'\theta}/u_*$,
$\overline{w'\theta}$	$-H/\rho_0 c_p$, where H is the flux of sensible heat,
L	Monin–Obukhov length = $-u_*^3 T_0/gk\overline{w'\theta_v}$,
$\overline{w'\theta_v}$	the flux of virtual potential temperature,
k	von Kármán's constant,
T_0 and ρ_0	reference absolute temperature and density for the surface layer,
g	acceleration of gravity,
c_p	specific heat of air at constant pressure.

In spite of careful field experiments that have been carried out at several locations around the world during the last 30 yr, there is still controversy regarding the exact form of the functions $\phi_m(z/L)$ and $\phi_h(z/L)$. During the 1960's, serious experimental attempts were made to resolve this question in Australia (several papers, see the summary by Dyer, 1974), in the U.S.S.R. (Zilitinkevich and Chalikov, 1968), and in the U.S.A. (the Kansas experiment, Businger *et al.*, 1971). Yaglom (1977) made a critical comparison of the results from these experiments as well as from some others, and found that the differences were appreciable. He came to the conclusion that '... it is quite probable that a considerable part of the scatter of the lines (in the plots of the various ϕ_m- and ϕ_h-expressions) is due to some experimental errors'. He was also concerned about the scarcity of data to verify the assumption of constancy of the turbulent fluxes τ and H with height.

Because of the suspicion that instrumental inadequacies was the cause of the differences, an international turbulence comparison experiment with participation of groups from five countries was arranged in Australia in 1976 (see Dyer *et al.*, 1982). The study indeed showed that instruments employed by the different groups gave significantly different results when various turbulence characteristics were compared. Various causes of error were identified and when corrections for these were made, the results agreed better. There were, however, appreciable discrepancies remaining. It was recognized that flow distortion can be an important source of error, which was taken into account by making the so-called 'tilt correction'. It was felt, however, that this might be inadequate; one of the conclusions of the paper was: 'Tilt equations appear to represent only a first approximation to the required correction'.

The dynamic nature of probe-induced flow distortion was elucidated in a theoretical study by Wyngaard (1981). The present author, Högström (1982), applied Wyngaard's results to careful wind tunnel measurements using a particular turbulence instrument developed by him. As shown in Högström (1982), the dynamical effects give rise to appreciable errors in the various calculated turbulence statistics – errors that are not remedied by the 'tilt correction procedure'. Attempts were made to redesign the instrument in order to remove the flow distortion. It was possible then to reduce the magnitude of the error but not to eliminate it entirely. It was, however, possible to derive an accurate empirical expression for the correction of the vertical component of the wind, which removes the error entirely.

It is to be expected that most turbulence instruments that have been employed in atmospheric surface-layer studies in the past to a larger or smaller extent suffer from dynamic flow distortion. The present author's results prompted Rasmussen *et al.* (1981) to carry out a similar study on their instrument, and they found indeed that large errors occurred. It is reasonable then to make the assumption that the systematically different results obtained by different groups in their respective determinations of ϕ_m and ϕ_h as functions of z/L may to a large extent be due to the failure of the various groups to correct properly for the dynamic flow distortion error. It is a logical step then to carry out a new, careful field experiment with a turbulence instrument that has been duly corrected for such errors. Results from such an experiment are presented in this paper.

2. General Description of the Measurements

2.1. The site

The measurements were carried out in May and June 1986 at a site, Lövsta, situated about 10 km southeast of the city of Uppsala (59° 50′ north) in southern Sweden. The area was chosen because it is particularly flat and has an unobstructed fetch of 1.5–5 km in a wide sector (see Figure 1). A 24 m tower was erected in a field planted with a mixture of clover and grass. This type of crop covered the area to the north-west of the road indicated in the figure. The area to the south-east of the road was planted with wheat. The actual height of the crop was measured every day during the experiment. When the measurements began on May 16, the height of both crops was 25 cm. It then gradually increased, reaching 75 cm on June 13. On the next day, the clover-grass crop on the north-west side of the road was cut, but the wheat crop was left growing during the remaining three days of the experiment.

The instrumented tower was placed about 10 m west of the road, and access to it was always by a narrow path running perpendicular to the road. In this way, the crop grew undisturbed right up to the tower for a large sector.

It can be seen from Figure 1 that the most favourable conditions for surface-layer measurements at this site are when the wind comes from the sector 220–310°. Then the fetch is absolutely uniform and unobstructed for 1.5 km or more. The original idea was to use only measurements with winds from this sector. Analysis of the measured data stratified according to wind direction showed, however, that instruments with winds from the sector 110–170° gave results that were statistically indistinguishable from those from 'the best sector', so data from the sector 110–170° have been included in the final analysis. We note that for this sector, the wind crossed the 7 m wide gravelled road at an angle between 80 and 130° and the unobstructed fetch was better than 1.5 km. When the wind was from the sector 170–220°, the data points were scattered and showed distinct irregularities as well, so measurements from this sector were

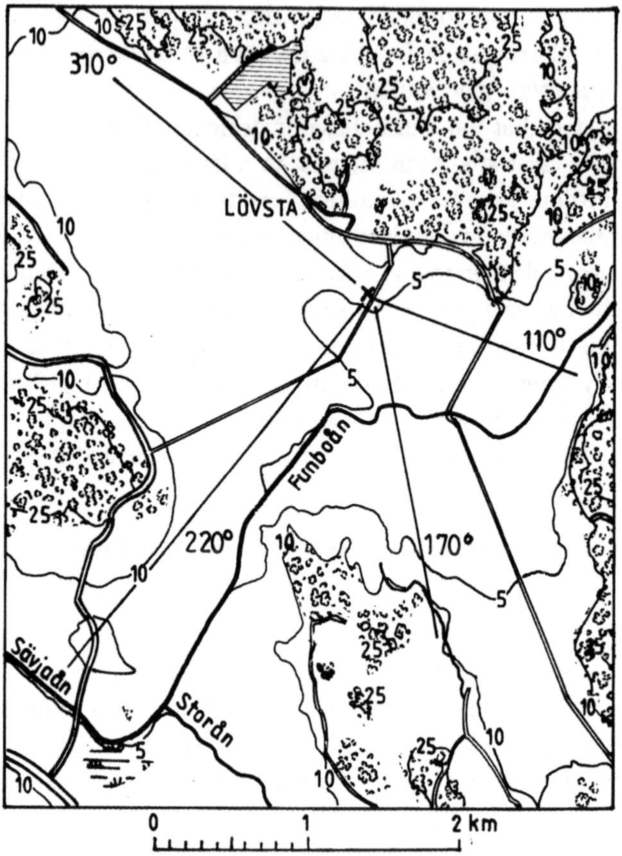

Fig. 1. Map of the experimental area.

excluded from the final analysis. Measurements from the remaining sector, 310–110° were also excluded, because of the limited fetch (< 1 km).

2.2. Instrumentation

The tower was equipped with slow-response 'profile' instrumentation for measuring wind speed, temperature and humidity at six levels and with fast-response 'turbulence'

TABLE I

Profile instrumentation during the Lövsta experiment. Measuring heights represent heights above the soil.

Measured property	Height in meters	Type of sensor
Wind speed	1.15, 2.07, 3.82, 6.96, 12.84, 23.97	Modif. Casella cup anemometers
Temperature	0.76, 1.68, 3.43, 6.57, 12.45, 23.58	Ventilated Pt500-sensors
Humidity	0.94, 1.86, 3.61, 6.75, 12.64, 23.76	Dry and wet bulb Pt-sensors
Wind direction	13, 24	Light wind vanes
Net radiation	ca. 1.5 m, on special stand	CSIRO instrument
Global radiation	ca. 2 m, on special stand	Moll Gorschinsky

instruments at 3, 6, and 13.9 m. The exact measuring heights for the profile instrumentation are given in Table I. All instruments were mounted on booms pointing from the tower towards the west. The tower is of open lattice triangular construction with 25 cm sides. The turbulence instruments were placed at the end of 125 cm booms. Thus tower interference is expected when the wind is from the sector 92–104°, which is well outside the sector used in the analysis. Table I also lists the types of sensors used in the profile instrumentation. The cup anemometers were calibrated in a wind tunnel prior to the experiment, the expected overall accuracy being ± 0.05 m s^{-1}. The temperature sensors used for 'dry' and 'wet bulb' measurements were calibrated in a water bath to within $+0.01$ °C for the difference measurements. Insulated and aspirated radiation shields of a design similar to that developed at the Cabauw site (Slob, 1978) were employed, so the accuracy of the temperature difference measurements in the field is expected to be no worse than ± 0.02 °C.

The turbulence instrument employed in this study is the MIUU instrument, described in Högström (1982). It is basically a wind vane-based three-axial hot-wire system

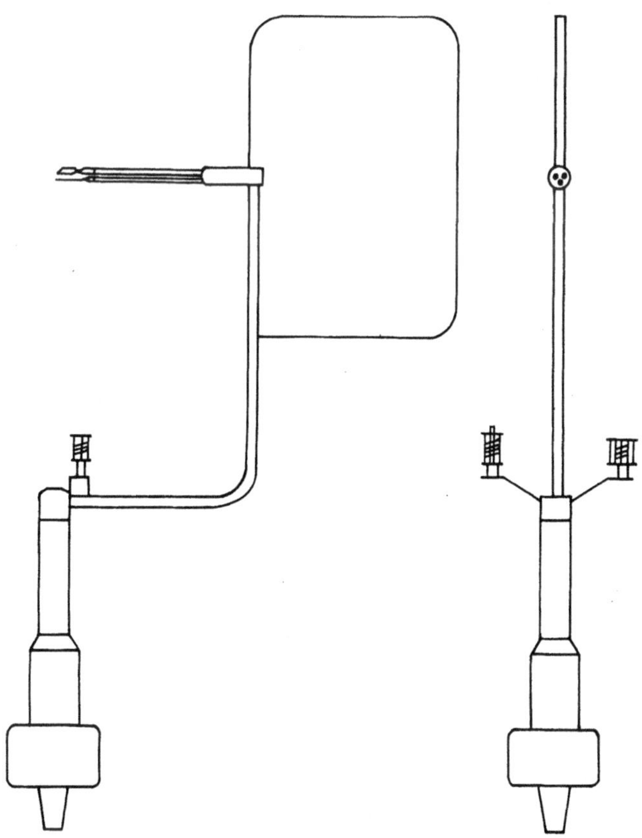

Fig. 2. The turbulence instrument (the MIUU instrument) used in the study. The instantaneous wind vector relative to the vane is measured with three 45° hot-film probes; the orientation of the vane relative to the fixed frame of reference is measured with a potentiometer. Dry- and wet-bulb sensors (25 μm Pt wire) are located ca. 20 cm below the hot-film probes, as shown in the figure.

(actually : hot film probes are used) supplemented with sensors for 'dry' and 'wet bulb' measurements; see Figure 2. A detailed discussion of the performance of the instrument is given in Section 3.

The surface-layer measurements were supplemented with Doppler sodar measurements on a continuous basis – to give information on wind and boundary-layer structure in the lowest 300 m – as well as tethered balloon soundings (wind, temperature, moisture) during the nights, and radio soundings with free flying balloons (Vaisala RS80 sondes) during most of the days.

2.3. The Experiment

Measurements were performed during altogether 96 hr scattered over the period May 16–June 17, 1986. The basic measuring period for the turbulence instruments was 100 min, with a sampling rate of 20 Hz for each of the signals: 3 instruments with 3 hot-film probes, 1 potentiometer, and two temperature sensors each. During two particular nights and mornings, the instrument was left in an unattended mode of operation, with a reduced sampling rate, 1 Hz.

The slow-response 'profile' instrumentation was sampled once per minute. The wind speed measurements were integrated over this time interval, so true averages were recorded. The temperature data were spot readings, but, because of the instrumental time constant (which is of the order 30 s), rapid temperature fluctuations were smoothed out.

The recordings were divided into 'runs' of 30 min duration, thus giving a basic data set of 192 runs. For the final analysis, the following criteria were, however, enforced.

(1) Wind direction from the sector 110–170° or from 220–310° (cf. p. 57);

(2) Reasonably steady-state conditions;

(3) For the ϕ_m-analysis: $u_* \geqq 0.1$ m s^{-1};

(4) For the ϕ_h-analysis: $|H| \geqq 10$ W m^{-2} and $u_* \geqq 0.1$ m s^{-1}.

Application of these criteria left 90 'runs' for the final analysis.

The first days of the experimental period were characterized by anomalously high wind speeds: 10–12 m s^{-1}. This situation in combination with relatively few clouds created numerous runs with near neutral stability. As seen in Section (6), this provides a good data basis for accurate determination of von Kármán's constant and the value of ϕ_h at neutrality. In June, the weather was dominated by a high pressure area, which gave rise to a much larger range in z/L.

3. Performance of the Turbulence Instrument

3.1. Pre-experimental Calibrations

The turbulence instrument used in this study is basically of the same design as that described in Högström (1982). Two important modifications have, however, been made subsequently: (1) The dry- and wet-bulb sensors have been modified and are now placed as shown in Figure 2 (for the reason for this modification, see Smedman and Lundin,

1986). (2) Instead of hot wires, which are very fragile, we now use fiber film sensors (Disa type 55R02). The geometry of these probes is exactly similar to that of the hot-wire probes except for the diameter of the sensor which is 70 μm in the case of the fiber film probe compared to 5 μm for the hot wires. Prior to the field experiment, we carried out an extensive wind tunnel test of the entire instrument equipped with these new probes. The main purpose of this study was to investigate in detail the directional sensitivity of these particular probes. This goal was successfully accomplished, and expressions for numerical evaluation of the true wind components from the simultaneous signals from the three hot-film sensors were derived, see Bergström and Högström (1987). As an important bi-product, we reproduced, with high accuracy, the expression for the flow distortion obtained many years earlier and reported in Högström (1982).

Prior to the field experiment, all three instruments with their individual hot-film sensors were calibrated in our own small calibration wind tunnel, in the way described in Högström (1982). The procedure was repeated several times to check the reproducibility and long-term stability of the measuring system. The result is very encouraging, the accuracy being $\pm 0.5\%$ over the entire wind speed range encountered in the calibration wind tunnel, 3–30 m s^{-1}, with no trend in the data over time. As seen below, there are strong reasons to believe that this result was indeed applicable during the field experiment as well.

3.2. Instrument Behaviour in the Field

As described in Högström (1982), the turbulence instrument is equipped with electronic tilt sensors which enable very accurate leveling of the instruments in the field. Then the measured mean values of the vertical velocity for each run is a sensitive test of the stability of the hot-film measurements, because, as a first approximation, we have for the average vertical velocity $\overline{w}$ for each run:

$$\overline{w} = \frac{1}{\sqrt{2}}(\overline{U}_A - \overline{U}_B),$$

where $\overline{U}_A$ and $\overline{U}_B$ are the run-average values of the horizontal wind component recorded by the two vertically oriented 45° slanting probes. For the Lövsta flat site, $\overline{w}$ should be close to zero. Thus even a slight instability in the measurement of each of the components U_A and U_B would show up in $\overline{w}$. As seen from the upper half of Table II,

TABLE II

Average vertical velocity (m s^{-1}) and average ratio $\overline{U}_{\text{prof}}/\overline{U}_{\text{turb}}$ for $\overline{u} \geqq 3$ m s^{-1}, where $\overline{U}_{\text{prof}}$ and $\overline{U}_{\text{turb}}$ are the horizontal mean wind speed measured with cup anemometers and turbulence instruments, respectively, at 3, 6, and 13.9 m. For each estimate the standard deviation of the individual run averages around the respective mean is also given.

	3 m	6 m	13.9 m
Vertical velocity	-0.017 ± 0.05	0.026 ± 0.04	0.010 ± 0.05
$\overline{U}_{\text{prof}}/\overline{U}_{\text{turb}}$	0.99 ± 0.02	0.99 ± 0.03	1.00 ± 0.03

$\overline{w}$ is indeed close to zero as an average for each of the instruments for the entire data set. Also, the individual run averages for $\overline{w}$ are small, the standard deviation of the individual values being only 4–5 cm s^{-1}.

Thus we conclude that the hot-film probes kept their calibrations with high accuracy throughout the experiment. The lower part of Table II displays a comparison of the horizontal wind speed measured with the turbulence instruments and with the cup anemometers (interpolation in logarithmic height to the exact level for the respective turbulence measurement). The quotient between 'the profile wind speed' and the 'turbulence wind speed' is seen to be close to unity, with a standard deviation of the individual values of 2–3% for $\overline{u} \geqq 3$ m s^{-1}. Below 3 m s^{-1} (not shown in the table), the ratio varies systematically with wind speed, being 1.07 for $\overline{u} < 1.5$ m s^{-1} and decreasing linearly to 1.00 at 3 m s^{-1}. This pattern appears at all levels and is believed to be due to deficiency of the wind tunnel calibrations at low wind speed. As discussed further in Section 5, the wind profiles are corrected for this systematic error.

3.3. Possible Systematic Errors

In the beginning of this section, we reported that the test in a large wind tunnel verified the empirical flow distortion correction introduced in Högström (1982). The analysis of the mean vertical wind measured in the field experiment not only verifies that the calibration of the hot-film probes was stable, but also supports the supposition that our flow distortion correction worked properly (although of course it is a necessary but not a sufficient condition that the computed mean vertical wind component is zero).

Other possible systematic errors relate to the effect of probe separation in the calculation of covariances. Koprov and Sokolov (1973) have studied this phenomenon experimentally. They found that the determination of $\overline{u'w'}$ is more susceptible to probe separation than the corresponding determination of $\overline{w'\theta}$. In our case, $\overline{u'w'}$ is likely to be entirely unaffected by this error, because the relevant sensors (the hot-film probes) are separated by less than 5 mm. The temperature sensor is, however, located 20 cm below the hot-film probes. Koprov and Sokolov suggest that the measured temperature flux $\overline{w\theta}$ is underestimated by the following factor:

$$f = \exp(-\Delta z/2z)\,, \tag{3}$$

where Δz is the vertical separation between the sensors for w and θ, and z is the height above the ground. For our measuring heights 3, 6, and 14 m, this would imply respectively: $f = 0.97$, 0.98, 0.99. As seen below, the mean of the measured fluxes at the three levels is taken to represent the flux at the ground. The above would then mean a possible underestimation of the flux by about 2%. The expression (3) is, however, only tentative, being based on measurements from just one level.

4. Check of the Constant-Flux Hypothesis

Table III displays statistics for τ and H as ratios of the values measured at 6 and 12 m relative to the corresponding value at 3 m. The results strongly support the constant-flux

TABLE III

Ratio of turbulent fluxes of momentum τ and of sensible heat H measured at 6 m and 13.9 m relative to the corresponding value at 3 m. For each estimate the standard deviation of the individual run ratios around the respective mean is also given.

τ_6/τ_3	τ_{14}/τ_3	H_6/H_3	H_{14}/H_3
0.97 ± 0.17	1.01 ± 0.22	1.05 ± 0.16	0.93 ± 0.15

hypothesis. Thus the ratios for the momentum flux is within $\pm 3\%$ of unity and the corresponding ratio for the sensible heat flux within $\pm 7\%$.

The table also gives the standard deviations of the corresponding ratios of the individual runs relative to the mean. It is relevant to compare the observed scatter with the scatter expected from the inadequate statistical sampling which arises when covariances are derived from a time series of 30 min duration. This comparison was carried out in Dyer *et al.* (1982), the result being that the statistical error in the determination of $\overline{u'w'}$ was $\pm 16\%$ and for $\overline{w'\theta'}$ $\pm 10\%$. Table III present *ratios* of the fluxes measured at two levels at the same time. There is the possibility then that the statistical errors are correlated to some degree. If this correlation were perfect, there would be no scatter in the flux ratios from this source, but at the other extreme, if the correlation were zero, the expected error in the ratios would be $\sqrt{2}$ times the values given by Dyer *et al.* This gives a theoretical span of 0– $\pm 0.32\%$ for the 'statistical scatter' in the τ-ratios and 0– $\pm 20\%$ for the H-ratios. Table III shows that the corresponding observed scatter is about $\pm 20\%$ for τ and $\pm 15\%$ for H, which is thus well within the corresponding theoretical ranges. As we cannot pin down the magnitude of the correlation between 'the statistical error' at the two levels, we cannot estimate the magnitude of the contributions to the scatter of the ratios due to other sources, notably from surface heterogeneities. No doubt, this contribution is of modest magnitude – probably less than $\pm 10\%$. More importantly, there is little indication of systematic effects of surface heterogeneities in the average flux ratios of Table III.

5. Calculation of the Dimensionless Profiles and the Monin–Obukhov Length

5.1. The Turbulent Fluxes and the Monin–Obukhov Length

The fluxes τ and H of Equations (1) and (2) are by definition the momentum flux and sensible heat flux *at the ground*. But as shown in the previous section, there was no systematic variation of the fluxes with height. Thus for each of the runs, we take the average of the measured fluxes at 3, 6, and 14 m as our basic flux quantities. An exception to this rule applies to many of the runs with stable stratification. Thus whenever it was found that the flux at 14 m was consistently (i.e., between successive 30 min runs) and significantly lower than at 3 and 6 m, the fluxes were determined as an average of the values measured at 3 and 6 m and the flux-gradient analysis was terminated at 6 m.

As noted in the Introduction, the Monin–Obukhov length used here:

$$L = \frac{-u_*^3 T_0}{kg\overline{w'\theta_v}}, \tag{4}$$

contains a water vapour buoyancy correction. As shown in Lumley and Panofsky (1964) we have approximately:

$$\overline{w'\theta_v'} = \overline{w'\theta}(1 + 0.07/\beta), \tag{5}$$

where β = the Bowen ratio = $H/E\lambda$

$E\lambda$ = the turbulent flux of latent heat.

The water vapour flux E was measured in all runs and at all three 'turbulence levels'. Humidity profiles were also measured, so an analysis of the corresponding ϕ_q-function could, in principle, have been performed. Problems with the frequency response of the wet-bulb sensor employed made, however, large corrections necessary, in particular for the fluxes measured at the lowest two levels (see Smedman and Lundin, 1986, for a discussion). Thus, rather than attempting this, we refrain from presenting an analysis of ϕ_q here. This uncertaunty is, however, immaterial because of the smallness of the term representing the water vapour flux correction to the Monin–Obukhov length, Equations (4) and (5). So for each run we use a best estimate of E obtained from the corrected maesurements, and we note that even an uncertainty as large as $\pm 25\%$ in E is not likely to result in errors in excess of $\pm 2\%$ in L.

Concerning the value of von Kármán's constant, we use 0.40, which is the value obtained from the ϕ_m-analysis for the near neutral range, as reported in Section 6.

Note, that the moisture correction (5) is applied in the definition of the Monin–Obukhov length only and not in the definition of ϕ_h, Equation (2), in keeping with the reasoning in Nicholls and Smith (1982).

5.2. The Gradients

As mentioned in Section 2, the height of the vegetation at the site was appreciable. This means that a zero-plane correction had to be applied in the gradient analysis, the effective height thus being $(z - d)$ rather than z. Several investigations (see, e.g., Baldocchi *et al.*, 1983) have shown that for a given crop there is close proportionality between vegetation height h and zero displacement d:

$$d = \alpha \cdot h\,. \tag{6}$$

The value of the factor α varies with the density and character of the crop. Thom (1971) presented a basic theoretical treatment of the problem which he tested using an artificial crop in a wind tunnel. He then carried out a very thorough field study, conducted in and above a bean crop, yielding $\alpha = 0.75$. As our crop is similar in density and rigidity to a bean crop, we adopted this value for α in the present analysis. At the end of this section, we shall present an analysis of the impact of a possible range of uncertainty in the determination of α.

The six measured 30-min averages of wind speed $\bar{u}$ and potential temperature $\bar{\theta}$ for each run were used to derive a second-order polynomial in $\ln(z - d)$:

$$\bar{u} = \overline{u_0} + A \ln(z - d) + B[\ln(z - d)]^2 , \tag{7}$$

where $\overline{u_0}$, A, and B were determined by the method of least squares, and with a corresponding expression for $\bar{\theta}$. All profiles determined in this way were computer plotted together with the measurements. Visual imspection of these plots revealed a good fit between regression curves and measurements in most cases.

Gradients of $\bar{u}$ and $\bar{\theta}$ were computed for three standard levels: z = 3, 6, and 14 m, from the derivatives of Equation (7) and the corresponding equation for $\bar{\theta}$ at these levels, respectively. The values of $\partial\bar{\theta}/\partial z$ obtained in this way were used, together with the corresponding values for u_* and H to derive values of ϕ_h with Equation (2). In this expression (as well as in the corresponding expression for ϕ_m, Equation (1)), von Kármán's constant was assigned a provisional value of 0.40, an hypothesis which is tested in Section 6. The corresponding values for $\partial\bar{u}/\partial z$ were corrected in the following way before they were entered into the expression for ϕ_m (Equation (1)):

In general we have

$$\bar{u}_m = \bar{u} \cdot f , \tag{8}$$

where $\bar{u}_m$ is the measured wind speed and $\bar{u}$ the 'true' wind speed. Comparison between cup anemometer measurements and the corresponding turbulence measurements presented in Section 3 suggest that for $\bar{u} > 3$ m s^{-1}, $f = 1$. For $\bar{u} \leqq 1.5$ m s^{-1}, the data indicate that $f = 1.07$. For the intermediate range, the measurements imply that f is a linear function of $\bar{u}$. Differentiating Equation (8) with respect to z gives:

$$\frac{\partial\bar{u}_m}{\partial z} = \frac{\partial\bar{u}}{\partial z} f + \bar{u} \frac{\mathrm{d}f}{\mathrm{d}\bar{u}} \frac{\partial\bar{u}}{\partial z} ,$$

so that

$$\frac{\partial\bar{u}}{\partial z} = \frac{1}{f + \bar{u}\dfrac{\mathrm{d}f}{\mathrm{d}\bar{u}}} \frac{\partial\bar{u}_m}{\partial z} = F \frac{\partial\bar{u}_m}{\partial z} . \tag{9}$$

From the measurements, $\mathrm{d}f/\mathrm{d}\bar{u} = -0.054$ for $1.5 < \bar{u} < 3.3$. Outside this range, $\mathrm{d}f/\mathrm{d}\bar{u} = 0$. Application of Equation (9) with the above expression for f shows that the correction factor F is 0.94 for $\bar{u} < 1.5$ m s^{-1}, increasing to a maximum of 1.08 at $\bar{u} = 2.3$ m s^{-1} and becoming unity for $\bar{u} > 3.3$ m s^{-1}.

We shall now investigate the effect of uncertainty in the determination of the factor α (Equation (6)) on the non-dimensional gradients ϕ_m and ϕ_h. We shall carry out the analysis for ϕ_m, but it is strictly similar for ϕ_h. When $d \neq 0$, Equation (1) must be written:

$$\phi_m = \frac{\partial \bar{u}}{\partial z} \frac{(z-d)k}{u_*}. \tag{1A}$$

Differentiation of Equation (7) gives:

$$\frac{\partial \bar{u}}{\partial z} = \frac{A}{z-d} + \frac{2B \ln(z-d)}{z-d},$$

so that

$$\phi_m = \frac{k}{u_*}[A + 2B \ln(z-d)]$$

or

$$\phi_{m_1}/\phi_{m_2} = [A_1 + 2B_1 \ln(z-d_1)]/[A_2 + 2B_2 \ln(z-d_2)],$$

where A_1, B_1 and A_2, B_2 have been determined from the same profile measurements but with different assumptions for the zero displacement d. The analysis shows that for a certain percentage error in α, the error increases with h (or, equivalently, with d). We thus start by looking at the case of maximum vegetation height observed, $h = 75$ cm and introduce a possible uncertainty range of $\pm 25\%$ in the parameter α of Equation (6). The result is not very sensitive to the particular data chosen for the analysis and gives the values shown in the first line of Table IV for the errors in the determination of ϕ_m. The lower line gives the magnitude of the error for a vegetation height of about 40 cm. We conclude that the uncertainty concerning the exact value of α is not likely to produce significant errors in the determination of ϕ_m and ϕ_h. Thus in the composite ϕ_m- or ϕ_h-plot as a function of $\zeta = (z-d)/L$, we do not expect a possible systematic error in excess of ± 1 to $\pm 2\%$, as the plot is made up of points representing the heights 3, 6, and 14 m in roughly equal proportions and vegetation heights from zero (newly cut crop) to 75 cm.

TABLE IV

Sensitivity test of dimensionless profiles on assumed uncertainty in the parameter $\alpha = d/h$, where d = zero-plane displacement height and h = vegetation height. The figures give percentage change of ϕ_m for a 25% change of α for the heights 3, 6, and 13.9 m. The first line represents the case of maximum vegetation height $h = 0.75$ m, the second line $h = 0.40$ m.

	3 m	6 m	13.9 m
$h = 0.75$ m	4.3	2.5	0.5
$h = 0.40$ m	2.8	1.6	0.3

6. Results

6.1. Determination of von Kármán's constant and the value of ϕ_h at neutrality

Figures 3a and 3b show plots of ϕ_m and ϕ_h, respectively, against $\zeta = (z - d)/L$ for the stability range $|\zeta| \leqq 0.1$. The curves drawn in the figures are the following second-order regressions:

$$\phi_m = 1.00 + 3.43\zeta + 8.4\zeta^2, \quad |\zeta| \leqq 0.1\,, \tag{10}$$

$$\phi_h = 0.95 + 5.24\zeta + 6.3\zeta^2, \quad |\zeta| \leqq 0.1\,. \tag{11}$$

Table V presents relevant statistics for these estimates. As stated in the previous section, we have provisionally set $k = 0.40$ in the calculation of ϕ_m and ϕ_h for the individual runs (Equations (1) and (2)). The regression equation for ϕ_m, Equation (10), strongly supports this value, giving

$$k = 0.40 \pm 0.007\,,$$

where 0.007 is derived from the standard deviation of the estimate of the intercept listed in Table V. We note that the value of von Kármán's constant obtained here is identical to that found in a previous study (Högström, 1985) based on a much more limited data set and from another site. In the previous section, the effect of a possible $\pm 25\%$ uncertainty in the zero displacement was discussed. It was shown that the net effect in the determination of ϕ_m was not likely to exceed $\pm 2\%$. This would then apply to the determination of von Kármán's constant as well. As this error and the error in the statistical determination are independent, it is fair to say that the probable net uncertainty in the determination of k is likely to be ± 0.01.

TABLE V

Regression analysis of $\phi_m = A_m + B_m\zeta + C_m\zeta^2$ and $\phi_h = A_h + B_h\zeta + C_h\zeta^2$ for $|\zeta| \leqq 0.1$ with standard errors of estimates

	A	B	C
ϕ_m	1.00 ± 0.018	3.43 ± 0.32	8.4 ± 5.9
ϕ_h	0.95 ± 0.039	5.24 ± 0.64	6.3 ± 11.8

For the value of ϕ_h at neutrality, Equation (11) and Table V give:

$$(\phi_h)_{\zeta=0} = 0.95 \pm 0.04\,.$$

Comparison of this value with corresponding results of other authors is not entirely straightforward, as the value of von Kármán's constant enters in the definition of ϕ_h. Thus Businger *et al.* (1971) got $(\phi_h)_{\zeta=0} = 0.74$, but they use $k = 0.35$. We postpone a discussion of this until we have presented the result for ϕ_m and ϕ_h over the full stability

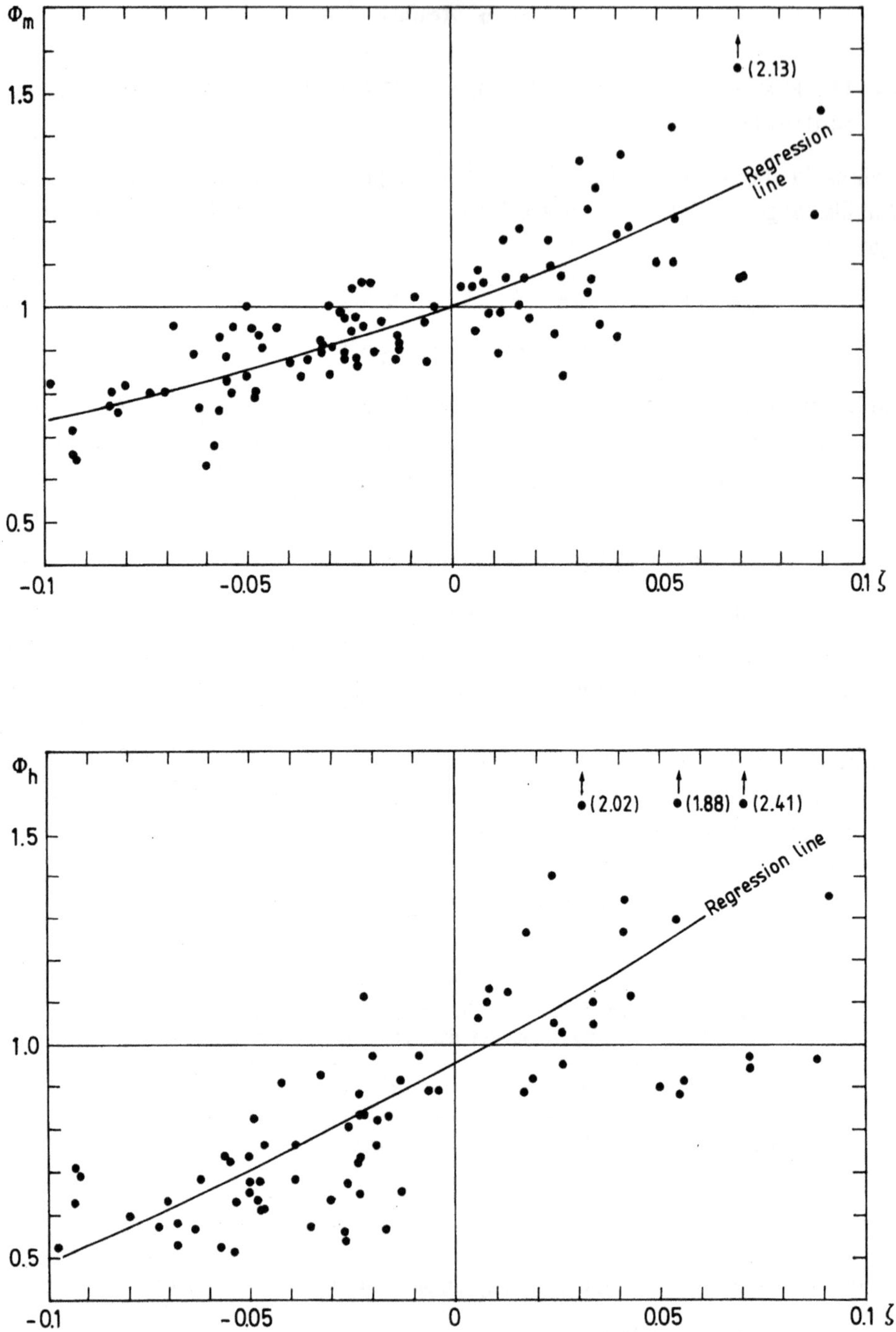

Fig. 3. Plots of ϕ_m and ϕ_h against $\zeta = (z - d)/L$ for $|\zeta| \leqq 0.1$.

range. Dyer and Hicks (1970) and later Dyer and Bradley (1982) obtained $(\phi_h)_{\zeta=0} = 1.0$, with $k = 0.41$ and $k = 0.40$, respectively, but both these studies were based on data from unstable stratification only, which makes it difficult to assess the exact value of ϕ_h at neutrality. For laboratory flow, according to Gibson and Launder (1978) '...the usually reported value...' of $(\phi_h)_{\zeta=0}$ is 0.93, which is remarkably close to our result: 0.95. We note that the probe separation error discussed in Section 3 is likely to reduce the value of ϕ_h by 1% or possibly 2%, which would bring our result in still closer agreement with the laboratory data.

From inspection of Table V, we also notice that the values given for the linear term in Equations (10) and (11) are likely to be accurate to within about $\pm 10\%$. This would imply that the slope of the ϕ_h-curve at neutrality differs significantly from the corresponding slope of the ϕ_m-curve there. We shall return to a discussion of this result later.

6.2. Analysis of ϕ_m and ϕ_h over an Extended Stability Range

The individual measurement points for ϕ_m and ϕ_h have been plotted on a logarithmic scale against the logarithm of $\zeta = (z - d)/L$ in four graphs: Figures 4 and 5 give ϕ_m and ϕ_h, respectively, for unstable stratification; Figures 6 and 7 for stable stratification. As

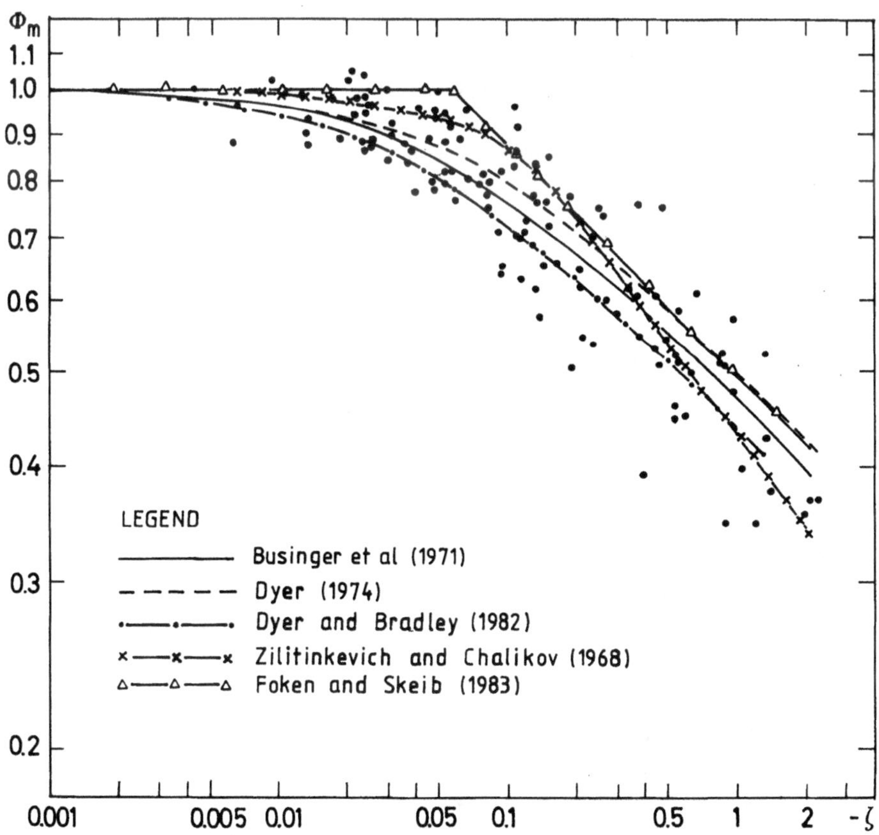

Fig. 4. Plot of ϕ_m against $\zeta = (z - d)/L$ in log-log representation for unstable stratification. The various curves represent modified expressions for ϕ_m, in accordance with Table VI.

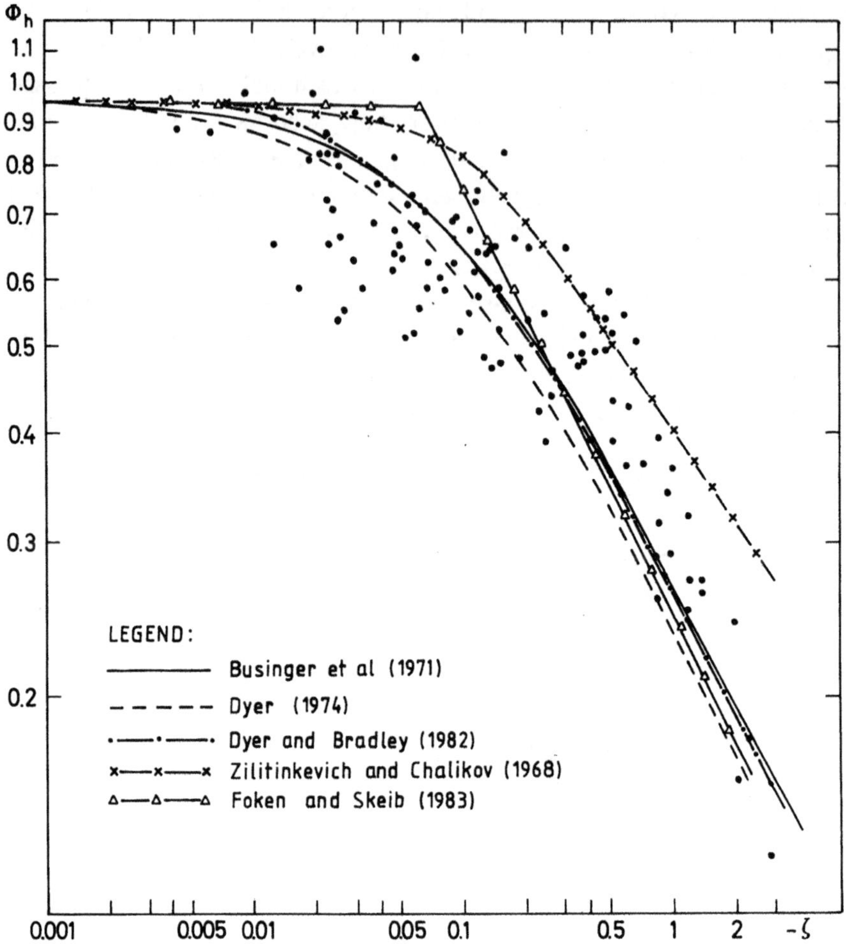

Fig. 5. As Figure 4, but for ϕ_h.

the scatter on the stable side is much larger than on the unstable side, stability group-means have also been entered in Figures 6 and 7 (triangles). The bars indicate the uncertainty of the estimate of the mean. The various curves in Figures 4 and 5 represent summaries of results from previous experiments as explained below.

Comparing different ϕ_m- and ϕ_h-formulations reported in the literature is not a straightforward matter. Yaglom (1977) in his comparative study recalculated all formulas to a common value for k, 0.4, and found large differences between the formulations. Here we shall attempt to go a step farther, the reasoning being as follows.

In the preceding sub-section, we showed that von Kármán's constant is likely to be close to 0.40. In a previous study, Högström (1985) not only obtained the same mean value for k, but he also concluded that k is actually constant (and not a function of the surface Rossby number, as had been suggested by Tennekes, 1973). It was also stressed

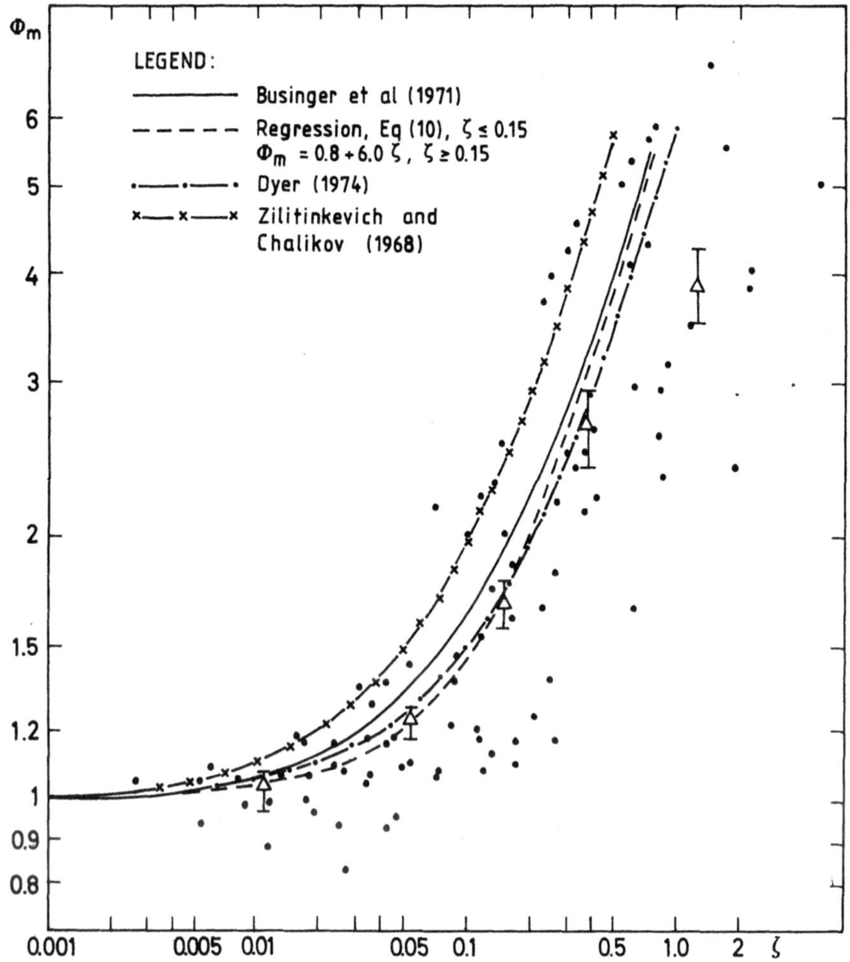

Fig. 6. Plot of ϕ_m against $\zeta = (z - d)/L$ in log-log representation for stable stratification. Triangles with error bars are stability group averages of the measurements with standard errors of estimates. The various curves represent expressions tested; cf. Table VI and the text.

in Högström (1985) that the value 0.40 for k is close to the result of a recent laboratory study by Purtell *et al.* (1981), which also showed that k is constant down to surprisingly low values of the Reynolds number. Thus it seems reasonable to assume that k is actually a constant, with a value close to 0.40. In consequence then, atmospheric studies that report k-values significantly different from 0.40 are likely to suffer from systematic measurement errors. As discussed in the Introduction, it is reasonable to suspect that such errors are due to dynamical flow distortion. Below we shall attempt to reevaluate the results from some of the most frequently quoted studied on the following premises: (1) the fluxes of momentum and sensible heat are multiplied by individual 'correction factors', so that the formulas for ϕ_m and ϕ_h give $\phi_m = 1$ and $\phi_h = 0.95$ at neutrality for $k = 0.40$; (2) these 'correction factors' are independent of stability (which might not be true, cf. Högström, 1982). Thus we can modify the original expressions for ϕ_m and ϕ_h

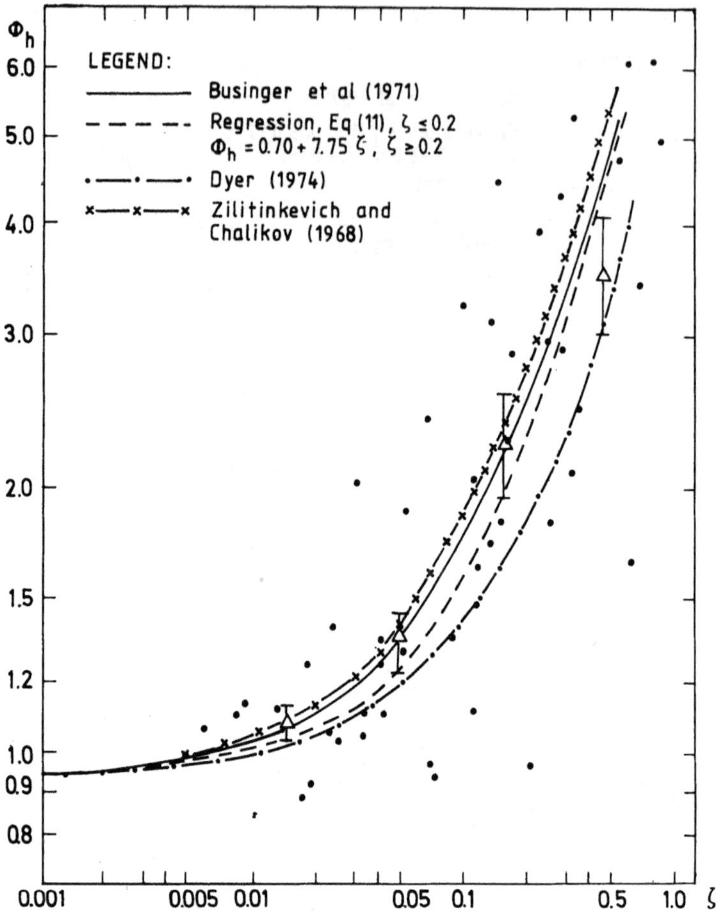

Fig. 7. As Figure 6, but for ϕ_h; cf. Table VII and the text.

and compare these modified expressions with one another and with the measurement results presented in Figures 4–7.

We shall illustrate the above procedure by showing in detail how the formulas of Businger *et al.* (1971) are modified. For $\zeta = 0$, they obtain:

$$\left.\begin{aligned}\phi_m &= 1\\ \phi_h &= 0.74\end{aligned}\right\} \quad \text{with} \quad k = 0.35\,.$$

Equations (1) and (2) then give, on our abovementioned premises:

$$\tau_K = u_{*K}^2 = 0.766 u_*^2 = 0.766\,\tau\,, \qquad (\overline{w'\theta})_K = 0.983\,\overline{w'\theta}\,,$$

where the index 'K' stands for 'Kansas' (and thus refers to the data of Businger *et al.*, 1971). The above also implies that the quantity $\zeta = z/L$ must be modified:

$$\zeta_K = 1.284\,\zeta\,.$$

Inserting the above expressions in the original Kansas formulas (cf. Tables VI and VII) gives the following modified expressions for unstable conditions:

$$\phi_m = (1 - 19\zeta)^{-1/4}, \tag{12}$$

$$\phi_h = 0.95\,(1 - 11.6\zeta)^{-1/2}. \tag{13}$$

These expressions which are represented by the solid curves in Figures 4 and 5, are seen to give a fair overall fit to our experimental data. For stable conditions, the modified Kansas expressions become:

$$\phi_m = 1 + 6.0\zeta, \tag{14}$$

$$\phi_h = 0.95 + 7.8\zeta. \tag{15}$$

These expressions are represented again by solid curves, in the corresponding graphs, Figures 6 and 7. It is seen that the ϕ_h-curve gives an excellent fit to the stability group-means at least up to $\zeta = 0.5$ where a possible levelling off begins, whereas the ϕ_m-curve is systematically high. Now, the regression equations for the near neutral range, Equations (10) and (11), give slopes at neutrality that are appreciably lower than indicated by Equations (14) and (15). A similar result was reported by Businger *et al.* (1971). Thus, there must be a region of ζ-values on the stable side for which the slope

TABLE VI

Formulas for ϕ_m. Original as well as modified expressions are given (see the text).

Authors	Comments	Formulas		
		A. Unstable stratification		
Businger *et al.*	$k = 0.35$	Original :	$\phi_m = (1 - 15\zeta)^{-1/4}$	
(1971)	$k = 0.40$	Modified:	$\phi_m = (1 - 19.3\zeta)^{-1/4}$	
Dyer (1974)	$k = 0.41$	Original :	$\phi_m = (1 - 16\zeta)^{-1/4}$	
	$k = 0.40$	Modified:	$\phi_m = (1 - 15.2\zeta)^{-1/4}$	
Dyer and Bradley	$k = 0.40$	Original :	$\phi_m = (1 - 28\zeta)^{-1/4}$	
(1982)	$k = 0.40$	Modified:	$\phi_m = (1 - 28\zeta)^{-1/4}$	
Zilitinkevich and	$k = 0.43$	Original :	$\phi_m = 1 + 1.45\zeta$	$-0.16 \leqq \zeta \leqq 0$
Chalikov (1968)			$\phi_m = 0.41(-\zeta)^{-1/3}$	$\zeta \leqq -0.16$
	$k = 0.40$	Modified:	$\phi_m = 1 + 1.38\zeta$	$-0.16 \leqq \zeta \leqq 0$
			$\phi_m = 0.42(-\zeta)^{-1/3}$	$\zeta \leqq -0.16$
Foken and Skeib	$k = 0.40$	Original :	$\phi_m = 1$	$\zeta \geqq -0.06$
(1983)			$\phi_m = (-\zeta/0.06)^{-1/4}$	$\zeta \leqq -0.06$
	$k = 0.40$	Modified:	same as above	
		B. Stable stratification		
Businger *et al.*	$k = 0.35$	Original :	$\phi_m = 1 + 4.7\zeta$	
(1971)	$k = 0.40$	Modified:	$\phi_m = 1 + 6.0\zeta$	
Dyer (1974)	$k = 0.41$	Original :	$\phi_m = 1 + 5\zeta$	
	$k = 0.40$	Modified:	$\phi_m = 1 + 4.8\zeta$	
Zilitinkevich and	$k = 0.43$	Original :	$\phi_m = 1 + 9.9\zeta$	
Chalikov (1968)	$k = 0.40$	Modified:	$\phi_m = 1 + 9.4\zeta$	

TABLE VII

Formulas for ϕ_h. Original as well as modified expressions are given (see the text).

Authors	Comments	Formulas		
		A. Unstable stratification		
Businger *et al.*	$k = 0.35$	Original :	$\phi_h = 0.74(1 - 9\zeta)^{-1/2}$	
(1971)	$k = 0.40$	Modified:	$\phi_h = 0.95(1 - 11.6\zeta)^{-1/2}$	
Dyer (1974)	$k = 0.41$	Original :	$\phi_h = (1 - 16\zeta)^{-1/2}$	
	$k = 0.40$	Modified:	$\phi_h = 0.95(1 - 15.2\zeta)^{-1/2}$	
Dyer and Bradley	$k = 0.40$	Original :	$\phi_h = (1 - 14\zeta)^{-1/2}$	$\zeta \leqq 0$
(1982)	$k = 0.40$	Modified[a]:	$\phi_h = (1 - 14)^{-1/2}$	$\zeta \leqq -0.01$
Zilitinkevich and	$k = 0.43$	Original :	$\phi_h = 1 + 1.45\zeta$	$-0.16 \leqq \zeta \leqq 0$
Chalikov (1968)			$\phi_h = 0.41(-\zeta)^{-1/3}$	$\zeta \leqq -0.16$
	$k = 0.40$	Modified:	$\phi_h = 0.95 + 1.31\zeta$	$-0.16 \leqq \zeta \leqq 0$
			$\phi_h = 0.40(-\zeta)^{-1/3}$	$\zeta \leqq -0.16$
Foken and Skeib	$k = 0.40$	Original :	$\phi_h = 1$	$\zeta > -0.06$
(1983)			$\phi_h = (-\zeta/0.06)^{-1/2}$	$\zeta \leqq -0.06$
	$k = 0.40$	Modified:	$\phi_h = 0.95$	$\zeta > -0.06$
			$\phi_h = (-\zeta/0.06)^{-1/2}$	$\zeta \leqq -0.06$
		B. Stable stratification		
Businger *et al.*	$k = 0.35$	Original :	$\phi_h = 0.74 + 4.7\zeta$	
(1971)	$k = 0.40$	Modified:	$\phi_h = 0.95 + 7.8\zeta$	
Dyer (1974)	$k = 0.41$	Original :	$\phi_h = 1 + 5\zeta$	
	$k = 0.40$	Modified:	$\phi_h = 0.95 + 4.5\zeta$	
Zilitinkevich and	$k = 0.43$	Original :	$\phi_h = 1 + 9.9\zeta$	
Chalikov (1968)	$k = 0.40$	Modified:	$\phi_h = 0.95 + 8.9\zeta$	

[a] Here we assume that the measurements are correct and that the formula suggested is valid for $\zeta \leqq -0.01$, but that the extrapolation to $\zeta = 0$ is not valid.

of the ϕ-curves changes rather rapidly. Our present data are not accurate enough to provide a conclusive description for this range. An attempt can, however, be made if we require that the regression equations (10) and (11) are valid up to a point $\zeta = \zeta_1$ where the slope is that of Equations (14) and (15), respectively. The result is then:

$$\begin{aligned} \phi_m &= 1 + 3.43\zeta + 8.4\zeta^2, \qquad \zeta \leqq 0.15, \\ \phi_m &= 0.80 + 6.0\zeta, \qquad \zeta \geqq 0.15, \end{aligned} \tag{16}$$

which is plotted as a dashed line in Figure 6 and

$$\begin{aligned} \phi_h &= 0.95 + 5.24\zeta + 6.3\zeta^2, \qquad \zeta \leqq 0.2, \\ \phi_h &= 0.70 + 7.75\zeta, \qquad \zeta \geqq 0.2, \end{aligned} \tag{17}$$

which is plotted in a corresponding way in Figure 7.

It is seen that the ϕ_m-representation, Equation (17), fits data better than does Equation (14), but that the corresponding expressions for ϕ_h, Equation (17), give a poorer result than the corresponding Equation (15). Here we must, however, bear in mind that the second-order terms of the regression equations (10) and (11) are very ill-defined. The conclusion may then be that the second-order term for the ϕ_m-curve is approximately correct, whereas that for the ϕ_h-curve should be much larger, thus giving a much more rapid approach to the slope given by Equation (15).

Tables VI and VII summarize several formulas for ϕ_m and ϕ_h, respectively, described in the literature. The selection of formulas is by no means complete. It includes the most often quoted formulations by Businger *et al.* (1971) and Dyer (1974) as well as that of Zilitinkevich and Chalikow (1968), which according to Yaglom (1977), is commonly used in the U.S.S.R. and other countries in eastern Europe. Also included are a recent analysis by Dyer and Bradley (1982) and an equally recent study by Foken and Skeib (1983). The tables give the original equations, k-values used and the equations that result after modification according to the procedure described above. The corresponding curves have been entered in Figures 4–7. Inspection of these tables and curves leads to the following conclusions:

For unstable stratification and ϕ_m, the Kansas curve is intermediate between the curves of Dyer (1974) and of Dyer and Bradley (1982). For $-\zeta \leqq 1$, the agreement between the Kansas curve and any of these curves is better than $\pm 10\%$. For ϕ_h, the curve of Dyer (1974) is slightly below the Kansas curve, whereas the curve of Dyer and Bradley (1982) is very close. Thus, generally speaking, for unstable conditions the curves of Businger *et al.* (1971), those of Dyer (1974) and those of Dyer and Bradley (1982) are in close agreement with one another and with the present data. Other formulations tested do not behave as well. Thus, the Zilitinkevich and Chalikov (1968) curve for ϕ_h gives values that are far too high, whereas their ϕ_m-curve comes closer to the Kansas curve and the present data points. Foken and Skeib (1983) argue that ϕ_m and ϕ_h have values of unity up to $-\zeta = 0.06$. As seen from Figures 4 and 5, this result is not substantiated by the present measurements. For $-\zeta$ slightly greater than 0.1, their result is, however, in fair agreement with the measurements and with the Kansas curve.

For stable stratification and ϕ_m, we find that Dyer's (1974) curve gives a good fit to the present data, whereas that of Zilitinkevich and Chalikov (1968) gives values that are too high throughout the stability range measured. For ϕ_h, the curve of Zilitinkevich and Chalikov (1968) gives a good fit, but Dyer's (1974) curve gives values that are too low compared to the measurements.

7. Conclusions

It has been demonstrated that the turbulence instrument used in the present study was properly corrected for dynamical flow distortion, was highly stable during the field experiment and was unlikely to be subject to other systematic sources of error. The profile measurements were also of high quality. A possible uncertainty concerning the zero-plane displacement has been shown not to introduce errors in excess of $\pm 2\%$ in the ϕ_m and ϕ_h estimates.

For all daytime runs and for some of the night-time runs as well, the turbulent fluxes were found to be virtually constant up to the highest measuring level, 14 m. For the more stable cases, the constant flux layer reached above the 6 m level but not above the 14 m level.

From the flux-profile analysis in the stability range $|\zeta| \leqq 0.1$, we conclude that von Kármán's constant $= 0.40 \pm 0.01$ and the value of ϕ_h at neutrality is 0.95 ± 0.04 (cf. below). The uncertainty ranges include possible effects from turbulence sensor separation and from the uncertainty in determination of zero displacement.

Several formulas for ϕ_m and ϕ_h described in the literature have been modified so as to comply with the above results at neutrality. Assuming further that the flux corrections so obtained are independent of stability gives modified formulas for ϕ_m and ϕ_h as function of ζ that can be compared with each other and with the present data. This comparison shows:

For unstable conditions, the modified Businger *et al.* (1971) curves for ϕ_m and ϕ_h fit the present data well and are in good agreement – better than $\pm 10\%$ for $-\zeta < 1$ – with the corresponding curves of Dyer (1974) and Dyer and Bradley (1982). The modified 'Kansas expressions' are:

$$\phi_m = (1 - 19.3\zeta)^{-1/4}$$

$$\phi_h = 0.95\,(1 - 11.6\zeta)^{-1/2}\,.$$

For stable conditions, the scatter of the present data is larger than for unstable conditions. Thus the conclusions are less firm. The tentative results are as follows. For ϕ_h, the modified Kansas curve

$$\phi_h = 0.95 + 8\zeta\,,$$

gives a good fit, at least for $\zeta \leqq 0.5$. The modified curve due to Zilitinkevich and Chalikov (1968) gives equally good agreement with the measurements. For ϕ_m, the modified Dyer (1974) curve

$$\phi_m = 1 + 4.8\zeta\,,$$

gives good overall fit to the present data for $\zeta \leqq 0.5$. The corresponding modified Kansas curve

$$\phi_m = 1 + 6\zeta\,,$$

gives values which are 0–18 % higher for this stability range. Because of the large scatter of the data points, it is, however, not possible to decide which of these expressions is to be preferred. For $\zeta > 0.5$, there are few data points, but there is an indication, at least in the ϕ_m plot, that the curve starts to level off somewhere in the range $0.5 < \zeta < 1$, in agreement with the commonly accepted results of Webb (1970).

In the above analysis it has been assumed that $\phi_h = 0.95$ at neutrality. This has some support in the laboratory data quoted by Gibson and Launder (1978), as noted earlier. From a physical point of view, it is, however, hard to find any reason for ϕ_h being different from ϕ_m at neutral stratification. The experimental uncertainty in the determination of ϕ_h at $z/L = 0$ is so large (± 0.04) that it might well accomodate the value 1.0 at a reasonable confidence level. With equally good fit to the data, the expressions for ϕ_h can then be slightly modified accordingly:

$$\phi_h = (1 - 12\zeta)^{-1/2}, \quad \zeta \leqq 0,$$

$$\phi_h = 1 + 7.8\zeta, \quad \zeta > 0.$$

Acknowledgements

The field experiment was made possible through a joint effort of the MIUU staff. I am particularly grateful to Dr Ann-Sofi Smedman, Mr Knut Lundin and to Mr Hans Bergström, who was also responsible for the computer processing of the data. The project was supported by a grant from the Swedish Natural Science Research Council (NFR Contract E-EG 2684–111).

References

Baldocchi, D. D., Verma, S. B., and Rosenberg, N. J.: 1983, 'Characteristics of Air Flow Above and Within Soybean Canopies', *Boundary-Layer Meteorol.* **25**, 43–54.

Bergström, H. and Högström, U.: 1978, 'Calibration of a Three Axial Fiber Film System for Meteorological Turbulence Measurements', *Dantec Information*, submitted.

Businger, J. A., Wyngaard, J. C., Izumi, Y., and Bradley, E. F.: 1971, 'Flux-Profile Relationships in the Atmospheric Surface Layer', *J. Atm. Sci.* **28**, 181–189.

Dyer, A. J.: 1974, 'A Review of Flux-Profile Relationships', *Boundary-Layer Meteorol.* **7**, 363–372.

Dyer, A. J. and Bradley, E. F.: 1982, 'An Alternative Analysis of Flux-Gradient Relationships at the 1976 ITCE', *Boundary-Layer Meteorol.* **22**, 3–19.

Dyer, A. J. and Hicks, B. B.: 1970, 'Flux-Gradient Relationships in the Constant Flux Layer', *Quart. J. Roy. Meteorol. Soc.* **96**, 715–721.

Dyer, A. J., Garratt, J. R., Francey, R. J., Mc Ilroy, I. C., Bacon, N. E., Hyson, P., Bradley, E. F., Denmead, O. T., Tsvang, L. R., Volkov, Y. A., Koprov, B. M., Elagina, L. G., Sahashi, K., Monji, N., Hanafusa, T., Tsukamoto, O., Frenzen, P., Hicks, B. B., Wesley, M., Miyake, M., and Shaw, W.: 1982, 'An International Turbulence Comparison Experiment (ITCE 1976)', *Boundary-Layer Meteorol.* **24**, 181–209.

Foken, Th. and Skeib, G.: 1983, 'Profile Measurements in the Atmospheric Near-Surface Layer and the Use of Suitable Universal Functions for the Determination of the Turbulent Energy Exchange', *Boundary-Layer Meteorol.* **25**, 55–62.

Gibson, M. M. and Launder, B. E.: 1978, 'Ground Effects on Pressure Fluctuations in the Atmospheric Boundary Layer', *J. Fluid Mech.* **86**, 491–511.

Högström, U.: 1982, 'A Critical Evaluation of the Aerodynamical Error of a Turbulence Instrument', *J. Appl. Meteorol.* **21**, 1838–1844.

Högström, U.: 1985, 'Von Kármán's Constant in Atmospheric Boundary Layer Flow: Re-evaluated', *J. Atm. Sci.* **42**, 263–270.

Koprov, B. M. and Sokolov, D. Yu.: 1973, 'Spatial Correlation Functions of Velocity and Temperature Components in the Surface Layer of the Atmosphere', *Izv. Atmospheric and Oceanic Physics* **9**, 178–182 (English translation pp. 95–98).

Lumley, J. L. and Panofsky, H. A.: 1964, *The Structure of Atmospheric Turbulence*, Interscience, New York, 239 pp.

Nicholls, S. and Smith, F. B.: 1982, 'On the Definition of the Flux of Sensible Heat', *Boundary-Layer Meteorol.* **24**, 121–127.

Purtell, L. P., Klebanoff, P. S., and Buckley, F. T.: 1981, 'Turbulent Boundary Layer Flow at Low Reynolds Number', *Phys. Fluids* **24**, 802–811.

Rasmussen, K. R., Larsen, S. E., and Jörgensen, F. E.: 1981, 'Study of Flow Deformation Around Wind-Vane Mounted Three-Dimensional Hot-Wire Probes', *Disa Information*, **26**, 27–34.

Slob, W. H.: 1978, *The Accuracy of Aspiration Thermometers*, KNMI Scientific Report W.R. 78–1, De Bilt, 26 pp.

Smedman, A. S. and Lundin, K.: 1986, 'Influence of Sensor Configuration on Measurements of Dry and Wet Bulb Temperature Fluctuations', *J. Atm. and Oceanic Tech.*, submitted.

Tennekes, H.: 1973, 'The Logarithmic Wind Profile', *J. Atm. Sci.* **30**, 234–238.

Thom, A. S.: 1971, 'Momentum Absorption by Vegetation', *Quart. J. Roy. Meteorol. Soc.* **97**, 414–428.

Webb, E.: 1970, 'Profile Relationships: The Lon-Linear Range and Extension to Strong Stability', *Quart. J. Roy. Meteorol. Soc.* **96**, 67–90.

Wyngaard, J. C.: 1981, 'The Effects of Probe-Induced Flow Distortion on Atmospheric Turbulence Measurements', *J. Appl. Meteorol.* **20**, 784–794.

Wyngaard, J. C., Coté, O. R., and Izumi, Y.: 1971, 'Local Free Convection, Similarity and the Budgets of Shear Stress and Heat Flux', *J. Atm. Sci.* **28**, 1171–1182.

Yaglom, A. M.: 1977, 'Comments on Wind and Temperature Flux-Profile Relationships', *Boundary-Layer Meteorol.* **11**, 89–102.

Zilitinkevich, S. S. and Chalikov, D. V.: 1968, 'Determining the Universal Wind-Velocity and Temperature Profiles in the Atmospheric Boundary Layer', *Izv. Atmospheric and Oceanic Physics* **4**, 294–302 (English version: pp. 165–170).

ON THE MEASUREMENT OF DRY DEPOSITION USING IMPERFECT SENSORS AND IN NON-IDEAL TERRAIN

B. B. HICKS and R. T. McMILLEN

NOAA/ERL Air Resources Laboratory, Atmospheric Turbulence and Diffusion Division, P.O. Box E, Oak Ridge, TN 37831, U.S.A.

(Received in final form 23 July, 1987)

Abstract. Important questions concerning the turbulent exchange of atmospheric pollutants between the air and natural surfaces urgently require answers, but sensors for many important species are not yet sufficiently well developed for use with standard micrometeorological methods. There is need, therefore, to develop methods by which deficient sensors can be used in micrometeorological applications. There is also need to extend micrometeorological methods to circumstances which do not satisfy the conventional perfect-site constraints. Here, methods based upon the assumption of cospectral similarity are explored. Initial tests indicate that it is possible to estimate daytime turbulent fluxes with sensors giving response times considerably greater than the values normally quoted for eddy correlation (e.g., 5 s instead of 1 s), and to compute first-order corrections for the error resulting from the lack of detection of high-frequency turbulence. It is suggested that a similar method might be used to derive flux data in terrain more complex than can be handled by conventional micrometeorology. The techniques outlined here should be applied only with caution, but appear adequate to permit the use of deficient sensors in some circumstances, and good sensors over some micrometeorologically deficient terrain.

1. Introduction

A major goal of micrometeorological research has been to develop a method for determining the atmosphere-surface exchange rates of properties that affect the behavior of the atmosphere, primarily momentum, heat, and water vapor. This focus has been broadened in recent years to include trace gas and particle exchange, as a result of steadily increasing concerns about air pollution and its consequences. In response to the need for information on exchange rates of trace chemical species, techniques developed earlier for studies of Reynolds stress and heat exchange (both sensible and latent) have been extended, wherever possible, to the cases of trace gases such as carbon dioxide, ozone, sulfur dioxide, and nitrogen oxides, and sometimes to particles. 'Dry deposition' can be considered a subset of the overall topic of atmosphere-surface chemical exchange, which is itself an extension of more classical considerations of turbulent fluxes. As discussed here, dry deposition is the turbulent exchange of trace gases and small particles from the atmosphere to the surface.

Recent progress in research on surface fluxes of trace chemical species is following similar lines of development as for past studies of heat and moisture exchange, although the pace is far more rapid. Conventional micrometeorological studies of turbulent exchange have needed to confront similar problems to those now being addressed in studies of dry deposition. For example, many innovative methods have been suggested to permit measurements to be made (even imprecisely) in demanding conditions. The so-called NIFTI technique (Hicks and Dyer, 1972; Dyer, 1975) is of special relevance

Boundary-Layer Meteorology **42** (1988) 79–94.

here, since it was intended to address a situation in which questions of both sensor limitations and imperfect exposure arose. The similarity methods addressed here are a further step along this line of development, occasioned by the urgent need to evaluate surface fluxes of trace chemical species in experimentally demanding situations.

In the following section, we shall attempt to clarify the difference between the experimental needs for intensive research studies and for monitoring. In subsequent sections, we shall address the experimental limitations imposed in the application of normal micrometeorological methods, and will propose methods based on cospectral similarity as a possible practical solution. The surface heat energy budget is then proposed as a means to avoid the need for a detailed micrometeorological component. Effects of signal noise are then examined in the context of noise levels associated with practical air pollution sensors. Some preliminary experimental results are presented, and relationships permitting canopy heat storage to be taken into account are given. Finally, we shall present results of a series of field trials designed to test the technique, using water vapor as a surrogate.

2. Requirements for Research and Monitoring

It is useful to differentiate between those techniques being developed for routine use in dry deposition monitoring activities, and those suitable for more limited application in research investigations of dry deposition processes. In the former category, emphasis is usually placed on the interpretation of air concentration data, while recognizing that such inferential methods are deficient; appropriate deposition velocities are not always available and fluxes are not always directed towards the surface. A basic goal of many research programs is to extend the present limited knowledge of deposition velocities (and the factors that control them) to circumstances and chemical species outside the range of current knowledge. Most of the techniques of deposition measurement included in the latter category of research methods are intended for use in investigations of the deposition velocity, rather than in programs to evaluate deposition fluxes routinely.

The immensity of the problem is obvious. Deposition velocities are not only species-specific, but also surface-dependent and time-varying. In essence, the deposition velocity problem can be viewed in terms of a three-dimensional array, ordered by the kind of pollutant, by the type of surface, and by the meteorological situation. The task is too great to produce direct experimental evidence corresponding to all available elements in such an array. Instead, the research programs are trying to provide coverage over most of the array and to minimize the need for unguided interpolation.

Hicks (1979) addresses the matter of pollutant sensing requirements for micrometeorological applications. Several distinctly different factors complicate the way in which micrometeorological gradient methods can be applied to measure dry deposition rates. First, gradient methods require the use of sensors capable of resolving differences with 1% precision, or better. This requirement is sufficiently demanding that field experiments employing gradient methods are relatively rare, and are often comparatively unproductive. Second, gradient methods usually require the specification of an appropriate turbulent diffusivity. Over simple surfaces (e.g., relatively smooth surfaces such as grass), diffusivities can be computed from supporting wind and temperature gradient measurements by applying well-accepted flux-gradient relations that account for effects of surface roughness and atmospheric stability. However, in many situations of steadily increasing practical importance (e.g., forest canopies) the applicability of these flux-gradient relations is open to question. Over trees, eddy diffusivities describing pollutant transfer cannot be assumed to be the same as for water vapor, heat, or momentum (see Garratt, 1978; Hicks *et al.*, 1979). The Bowen ratio approach (e.g., Kanemasu *et al.*, 1979) is often used as a practical method to bypass this particular problem; by relating all measurements to a flux which can be measured with confidence (the net radiation), the need for a precise determination of eddy diffusivity is eliminated. However, further difficulties then arise with the need to account for heat storage in the canopy, as will be discussed later.

Eddy correlation techniques do not require an assumption regarding eddy diffusivity, and hence tend to be preferred. However, the fast-response sensing requirement limits the application to very few chemical species. In general, $\pm 10\%$ evaluations of eddy fluxes require 1 s response time in daytime conditions, and sometimes as short as 0.1 s at night. For most pollutant species, deposition occurs mainly in daytime and hence the requirement communicated by micrometeorologists to those developing new chemical sensors most often concerns the need for response times of one second or shorter.

Eddy correlation appears to be preferred whenever suitable instrumentation is available. However, gradient methods provide the only available micrometeorological alternative for many pollutants which cannot yet be measured rapidly enough for eddy correlation, such as nitric acid vapor (Huebert and Roberts, 1985).

The history of development of micrometeorological measurement methods has cast increasing attention on studies of the nature of turbulent exchange, rather than on the exchange (i.e., the flux) itself. Studies of the nature of atmospheric turbulence require that experiments be performed in near-laboratory conditions, leading necessarily to the school of 'flat-earth, uniform-surface, steady-state micrometeorology' that tightly focuses attention on the need to derive answers to basic questions that can only be addressed in such circumstances. It must be remembered, however, that one reason for conducting such studies is the need to develop and demonstrate techniques to make both measurements and predictions in far less ideal circumstances, more typical of natural landscapes. It is such surfaces which constitute the most important target for studies of trace gas and particle exchange, and hence it is imperative to step beyond the constraints of laboratory-philosophy micrometeorology.

3. Co-spectral Similarity

The essence of the suggested approach is the assumption that the ratio of two fluxes can be closely approximated by the ratio of two characteristics of the corresponding turbulence fields. This is likely to be an acceptable assumption whenever the net transfer of the two quantities in question is accomplished by the same field of eddy motions.

Thus, in concept it is required that the two fluxes be of substances with the same distributions of sources (or sinks) at the surface, or that measurements of the turbulence be made sufficiently far above the surface that any source-sink inequalities become inconsequential.

The theory of co-spectral similarity has been discussed thoroughly elsewhere (Wyngaard and Coté, 1972). Here, prime attention will be on practical applications of the theory, rather than on the theory itself. However, special effort will be made to emphasize limitations, such as the difficulties arising over forests. It is assumed that standard siting requirements are met, concerning horizontal homogeneity, adequate fetch, and stationarity. Later discussion will address how some of these requirements may perhaps be relaxed.

Kaimal *et al.* (1972) summarize spectral data in terms of normalized frequency and atmospheric stability, for temperature and velocity components. The normalization of frequency involves the height of measurement z and the wind speed u. The height scale of direct relevance is that above the level of the appropriate zero plane, and thus need not necessarily be the same for all of the quantities of interest. For example, we might expect to find that the eddies transporting radon emitted from the soil beneath a canopy are more deeply penetrating (hence larger and slower) than those associated with momentum transfer (for which the effective level of action is usually in the upper part of the canopy). Such matters are discussed at length elsewhere (see Garratt, 1978; Hicks *et al.*, 1979; Hicks, 1985); for the present, it is sufficient to recognize that there is no *a priori* reason that all eddy transfer should be characterized by the same effective zero plane displacement levels. In the present context, care should be taken to ensure that potential errors are small.

Similar conceptual difficulties arise over surfaces that are irregularly vegetated or topographically complicated. In this case, inequalities in the horizontal distributions of effective sources and sinks will also cause transport by different eddies; in addition, streamlines will be affected by topographic relief in a manner that greatly complicates the direct application of standard eddy correlation or gradient techniques.

4. Method Development

In general, the surface heat energy budget can be written as

$$H + LE = R_n - S\,, \tag{1}$$

where H is the sensible heat flux, LE is the latent heat flux, and S is the rate of heat storage beneath the level of measurement. The Bowen ratio, β, is defined as H/LE, so that

$$H(1 + 1/\beta) = R_n - S\,. \tag{2}$$

Suppose that a measure of pollutant concentration standard deviation σ_c is measured in some selected frequency band, and that similarly constrained (i.e., defined by the same frequencies) measures of temperature and humidity standard deviations are also obtained: σ_q and σ_T, respectively. If the flux of pollutant is F, then similarity principles dictate that

$$F/H \cong \sigma_c/c_p\sigma_T \tag{3}$$

and

$$H/LE = \beta = c_p\sigma_T/L\sigma_q\,. \tag{4}$$

From (4), trivial mathematical manipulation yields that

$$F \cong \sigma_c(R_n - S)/(c_p\sigma_T + L\sigma_q)\,. \tag{5}$$

Equation (4) provides a means for evaluating β without measuring gradients or covariances. Equation (5) yields an approximation of the flux F from imperfect (but matched) standard deviation data σ_c, σ_T, and σ_q, using measurements of net radiation as a basis. Net radiation can be measured using relatively simple instrumentation, without requiring anemometry or fast-response sensors. In daytime, the heat storage term S is typically less than 15% of R_n, and in most circumstances can be taken to be about 10% of R_n. Thus, (5) can be approximated as

$$v_d \cong 0.9(\sigma_c/C)R_n/(c_p\sigma_T + L\sigma_q)\,, \tag{6a}$$

where the deposition velocity v_d is defined as $-F/C$. (Note that the sign convention used here follows micrometeorological convention, so that turbulent fluxes are positive upwards.)

A somewhat simpler form of (6a) might be of special interest in circumstances in which the Bowen ratio β can be computed from external information, such as by use of a combination formula derived from agricultural or agrometeorological considerations. In this case, the need for specific humidity data to yield σ_q can be avoided, since (6a) can be rewritten as

$$v_d \cong 0.9(R_n/\overline{C})\,(\sigma_c/c_p\sigma_T)/(1 + 1/\beta)\,. \tag{6b}$$

The development above makes use of the equally-constrained turbulence-intensity standard deviations σ_T, σ_q, and σ_c as basic indicators of the corresponding fluxes of heat, moisture, and species mass. In practice, any of several indicators may be suitable, such as vertical differences ΔT, Δq, and ΔC over identical height intervals Δz. However, experience in initial field tests indicates that there is considerable benefit in using covariances with velocity components as indicator quantities. Work conducted so far has made use of vertical velocity to construct band-pass covariances C_{wT}, C_{wq}, and C_{wc}, which are used by direct substitution for σ_T, σ_q, and σ_c in the development given above, with no loss of generality:

$$v_d \cong 0.9(C_{wc}/C)R_n/(c_pC_{wT} + LC_{wq})\,, \tag{7a}$$

$$v_d \cong 0.9(R_n/C)\,(\overline{C}_{wc}/C_pC_{wT})/(1 + 1/\beta)\,. \tag{7b}$$

The revised relationships offer several advantages:

– The fundamental noise-rejection capabilities of covariance computations are used to advantage. This is of special importance in the case of a pollutant sensor (e.g., of SO_2, O_3, or NO_2), since unavoidable high-frequency noise contaminates all field measurements of σ_c regardless of the frequency band that is selected.

– The revised relations re-introduce the capability to differentiate between upward and downward fluxes.

– A simple band-pass filter can be applied to the shared signal used to compute the covariances. This helps ensure that the covariance indices are computed similarly, as is required by the underlying theory.

It should be noted that there is no requirement for either precise alignment of velocity sensors or for special analytical treatment to compensate for related measurement errors. The implications of this apparent lack of strong sensitivity to sensor orientation remain to be fully explored. For the present, it is sufficient to recognize that the method offers promise of some relaxation of stringent site and orientation requirements usually associated with micrometeorological field studies.

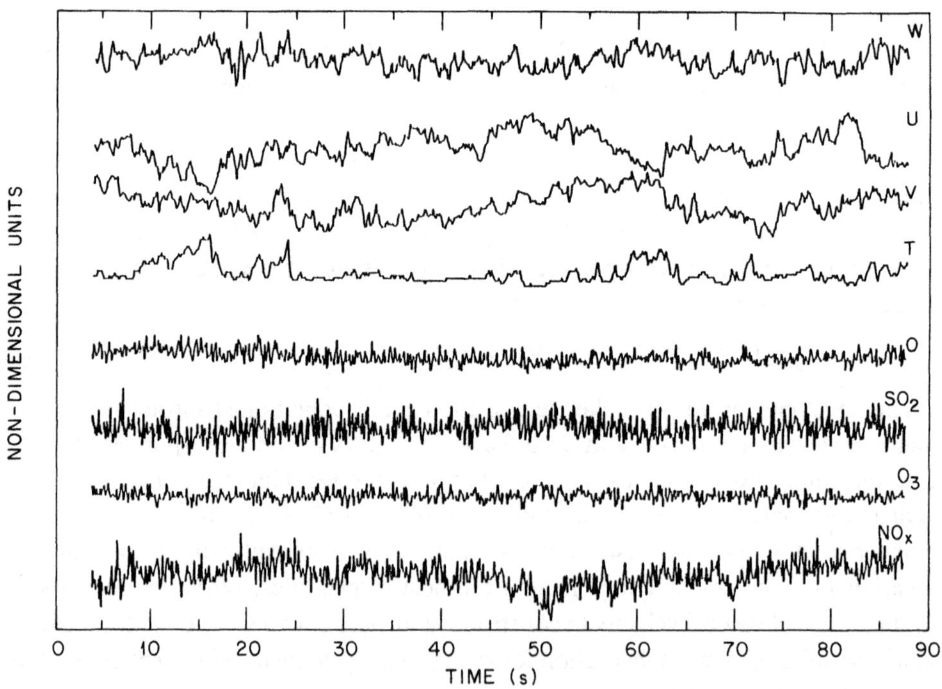

Fig. 1. An example of fast-response sensor outputs obtained during a recent field study conducted near State College, PA. All signals have mean values removed, using a 300 s recursive digital filter. The upper three traces are of wind velocity components: vertical, longitudinal and lateral. The fourth trace is of temperature, and the fifth represents humidity. The lowest three traces are of concentrations of sulfur dioxide, ozone, and nitrogen dioxide.

5. Spectra and Noise

Figure 1 is an illustration of records obtained using fast-response meteorological and air quality sensors during a recent study of dry deposition to agricultural crops, at State College, Pennsylvania. The traces demonstrate the high frequency noise that is a characteristic of all trace gas and aerosol sensors when operated near their limits of detectability. The speed of response required for normal day-time eddy correlation applications is about one second. For application in the way considered here, considerably slower time response may be adequate, as will be demonstrated later.

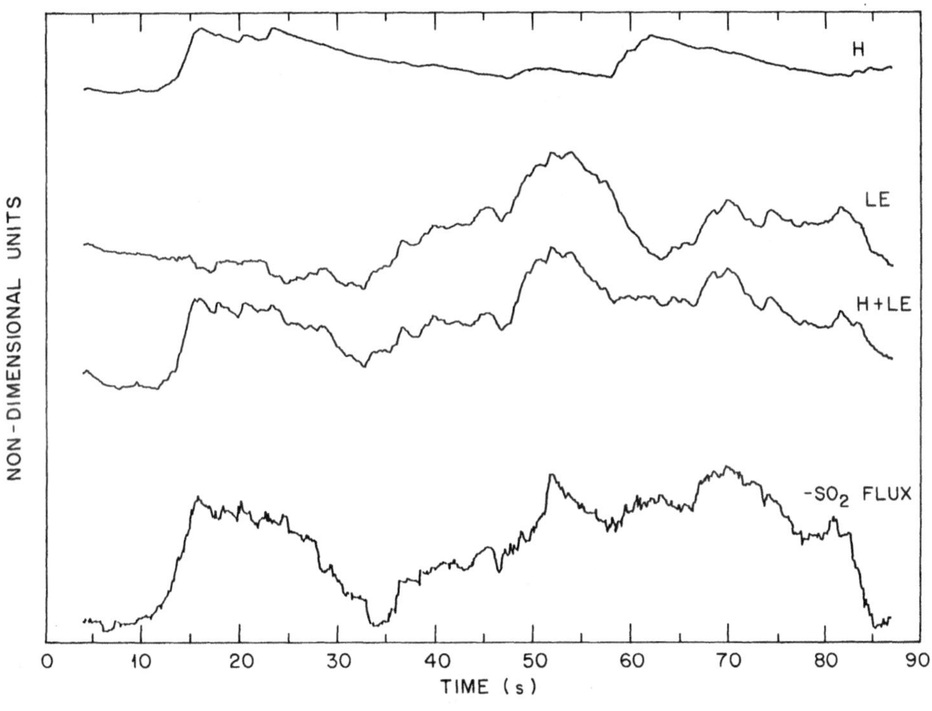

Fig. 2. 'Instantaneous' fluxes inferred from the fast-response traces shown in Figure 1, for sensible heat (H), latent heat (LE), and SO_2 exchange. Fluxes are computed as average covariances (using the vertical velocity), averaged over 20-s running mean. Note the close correspondence between the SO_2 exchange and the total turbulent heat flux $H + LE$.

Figure 2 demonstrates the ability of correlation techniques to extract meaningful signals from noisy data. The bottom curve is of running-mean covariances computed using the vertical velocity data of Figure 1 and the SO_2 concentrations. It is usually the time average (typically over 30-min periods) of such data that is of main interest. For the present, it is instructive to note the strong correlation between the SO_2 covariance record and that of sensible heat (H) and latent heat (LE) derived similarly (as smoothed covariances with temperature and humidity, respectively), and especially the strong correspondence with the total turbulent heat exchange, $H + LE$. It is this similarity of

TABLE I

Examples of data generated by a test of the theory of similarity methods for deducing pollutant deposition from radiation data and heat budget considerations, using slow-responding pollution sensors. All data are for 30-min runs, above a forest canopy. Data are quoted for 1, 3, 10, and 30 s response time sensors. Units for global (R_g) net radiation (R_n), sensible heat flux (H) and latent heat flux (LE) are W m^{-2}. All other quantities are normalized to unity for 'perfect' sensors. Fluxes F1, F2, and F3 are of SO_2, small particles (0.5–0.7 μm diameter), and larger particles (0.7–1.2 μm), respectively. The remaining quantities are standard deviations, also normalized, and in the same sequence.

	H	LE	F_1	F_2	F_3	ST	S_q	S_1	S_2	S_3
1500–1530, 2 May, 1983, R_g = 775, R_n = 540										
	224	279	1.00	1.00	1.00	1.00	1.00	1.00	1.00	1.00
1 s	220	274	1.00	1.00	1.00	0.99	0.99	1.00	0.99	0.98
3 s	205	259	0.99	1.00	1.00	0.98	0.98	1.00	0.99	0.95
10 s	169	228	0.97	0.94	0.92	0.94	0.94	0.99	0.97	0.91
30 s	71	149	0.90	0.84	0.80	0.85	0.86	0.96	0.95	0.87
1530–1600, 2 May, 1983, R_g = 640, R_n = 465										
	245	205	1.00	1.00	1.00	1.00	1.00	1.00	1.00	1.00
1 s	242	202	0.99	1.00	1.03	0.98	0.99	0.99	0.94	0.94
3 s	231	192	0.97	0.83	1.27	0.96	0.96	0.97	0.85	0.85
10 s	197	160	0.92	0.83	0.92	0.91	0.92	0.95	0.69	0.72
30 s	151	119	0.77	0.83	0.89	0.84	0.84	0.90	0.58	0.59
1600–1630, 2 May, 1983, R_g = 576, R_n = 385										
	166	124	1.00	1.00	1.00	1.00	1.00	1.00	1.00	1.00
1 s	162	120	1.00	0.98	0.92	0.98	0.99	0.94	0.94	0.94
3 s	147	108	1.01	0.84	0.60	0.96	0.96	0.88	0.84	0.84
10 s	108	78	0.88	0.66	0.27	0.91	0.91	0.78	0.66	0.69
30 s	62	44	0.33	0.79	0.21	0.82	0.82	0.70	0.51	0.56
1630–1700, 2 May, 1983, R_g = 473, R_n = 302										
	136	92	1.00	1.00	1.00	1.00	1.00	1.00	1.00	1.00
1 s	132	88	0.91	0.89	0.99	0.98	0.98	0.94	0.89	0.95
3 s	121	80	0.71	1.20	0.87	0.96	0.96	0.89	0.84	0.85
10 s	95	60	0.63	0.21	0.63	0.90	0.90	0.81	0.69	0.72
30 s	51	31	0.15	1.01	0.48	0.81	0.81	0.71	0.52	0.56
1700–1730, 2 May 1983, R_g = 352, R_n = 196										
	109	82	1.00	1.00	1.00	1.00	1.00	1.00	1.00	1.00
1 s	108	80	0.99	0.74	0.83	0.98	0.99	0.95	0.94	0.94
3 s	99	74	0.95	1.02	0.65	0.96	0.96	0.89	0.85	0.83
10 s	79	59	0.73	0.38	0.99	0.91	0.90	0.78	0.68	0.69
30 s	47	39	0.29	−0.32	1.38	0.67	0.83	0.67	0.52	0.54
1730–1800, 2 May 1983, R_g = 243, R_n = 103										
	125	49	1.00	1.00	1.00	1.00	1.00	1.00	1.00	1.00
1 s	123	47	0.99	0.90	1.10	0.98	0.96	0.93	0.94	0.94
3 s	114	43	0.96	0.96	0.87	0.96	0.94	0.87	0.84	0.85
10 s	98	36	0.82	0.19	1.37	0.91	0.89	0.75	0.67	0.69
30 s	55	21	0.57	0.30	1.10	0.82	0.79	0.62	0.50	0.53

turbulent exchange processes that underlies the techniques suggested here, and it is the inbuilt noise-rejection capability of eddy-correlation that permits the approach to be applied in the case of noisy pollutant sensors.

Table I presents an example of results obtained in a field test conducted at the Walker Branch Watershed forest meteorology field site on the Department of Energy's Oak Ridge (Tennessee) reservation. The data were obtained during a multilaboratory 'CORE' research experiment (see Hales *et al.*, 1987) conducted in collaboration with Argonne National Laboratory, Pennsylvania State University and the EPA Atmospheric Sciences Research Laboratory. The tabulation includes information on temperature, humidity, atmospheric sulfur concentration, and particle concentrations in two size ranges, approximately 0.5–0.7 and 0.7–1.2 μm diameter. The effect of using pollutant sensors with relatively slow response times has been simulated by smoothing records obtained with fast-response sensors to produce 1, 3, 10, and 30-s averages prior to computation.

The simulation demonstrates the expected noise reduction achieved by data averaging. The smoothing accomplished by signal averaging, often by application of capacitive feedback in electronic circuits, is a major reason that monitoring instruments are normally designed to have relatively slow response (i.e., minutes rather than seconds).

6. Heat Storage

The term S in Equations (1), (2), and (5) has so far been approximated as $0.1R_n$. The matter of heat storage and the evaluation of S has been investigated at the Walker Branch Watershed following similar procedures to earlier studies reported by Hicks *et al.* (1975) and Baldocchi *et al.* (1985). In the present instance, the Walker Branch field site is seen as an extreme test of the importance of S, since the site is forested and independent measurement of S in such circumstances is difficult. In practice, S is made up of several terms, primarily the heat storage S_1 involved in the biomass:

$$S_1 = Mc_b(\partial T_b/\partial t)\,, \tag{8}$$

where M is the quantity of biomass per unit surface area, c_b is the appropriate specific heat, and $\partial T_b/\partial t$ is the rate of change of biomass temperature T_b with time t. Heat storage associated with the air mass below the level of measurement (S_2) is also likely to be important, especially when foliage is sufficiently dense that subcanopy air temperatures rise to very high levels:

$$S_2 \simeq \rho c_p h(\partial T_a/\partial t)\,, \tag{9}$$

where h is the height of measurement of H and LE, T_a is the temperature of the air beneath the level h, and ρ is air density. The heat storage associated with changes of specific humidity (S_3) need also to be considered:

$$S_3 = h(\rho c_w)T_a(\partial q/\partial t)\,, \tag{10}$$

where c_w is the specific heat of water vapor.

A final term, G, concerns heat conduction directly into the ground. G is typically small in circumstances of a dense vegetative cover (such as at the Walker Branch field site, where the canopy is a 25-m oak-hickory forest), but for crops, grassland, and bare soil, G is far more important than either of S_1, S_2, or S_3.

It is possible to summarize the likely magnitudes of the various terms contributing to the net storage term S (as is evident in the present diagrams, and as discussed by Thom, 1975, and Baldocchi *et al.*, 1985), or to compute them using detailed sub-canopy measurements or models (e.g., McCaughey, 1985). Table II is intended to provide an indication of the relative magnitudes for sunny, summer conditions.

TABLE II

Estimates of heat storage terms contributing to the surface energy balance for measurements made at a height of 5 m above the vegetation in different circumstances. All values are expressed as a proportion of the available net radiation, R_n.

	Biomass storage S_1	Air thermal storage S_2	Air humidity storage S_3	Ground heat conduction G
Forest	10%	1%	1%	1%
Corn	4%	0	0	6%
Pasture	0	0	0	10%
Bare soil	0	0	0	15%

Sources: Baldocchi *et al.* (1985); Hicks *et al.* (1983); Hicks *et al.* (1972); Pruitt *et al.* (1973); Dyer and Hicks (1970).

Figure 3 demonstrates the possible consequences of neglecting S, or alternatively of applying the first-order correction represented by the assumption that $S \simeq 0.1R_n$, as in Equations (6) and (7). The data contributing to Figure 3 enable direct evaluation of S in Equation (1). Figure 4 relates the term S to the rate of change of canopy temperature with time $\partial T_b/\partial t$. There is considerable uncertainty associated with the determination of S, since it is computed as a small residual among several imperfectly-measured quantities (especially H and LE). As discussed elsewhere (Hicks *et al.*, 1983), Figure 4 does not show all the data; averages that are not significantly different from zero are omitted, since they do not contribute significantly to elucidating the underlying physics.

The biomass temperature data used to obtain Figures 3 and 4 were intensive, repetitive measurements using a hand-held infrared thermometer at many levels in the canopy. Canopy temperature profiles derived using these data have been presented elsewhere (Hicks, 1985). In practice, such direct measurements of biomass temperature are rarely available, and in-canopy air temperatures are typically used instead (Baldocchi *et al.* (1985) present a relevant discussion regarding the same site). It appears likely that such air temperature data would be sufficiently accurate indicators of biomass temperature for the present purposes, although doubtlessly some errors will arise.

The validity of the interpretations given above is illustrated by the line drawn through

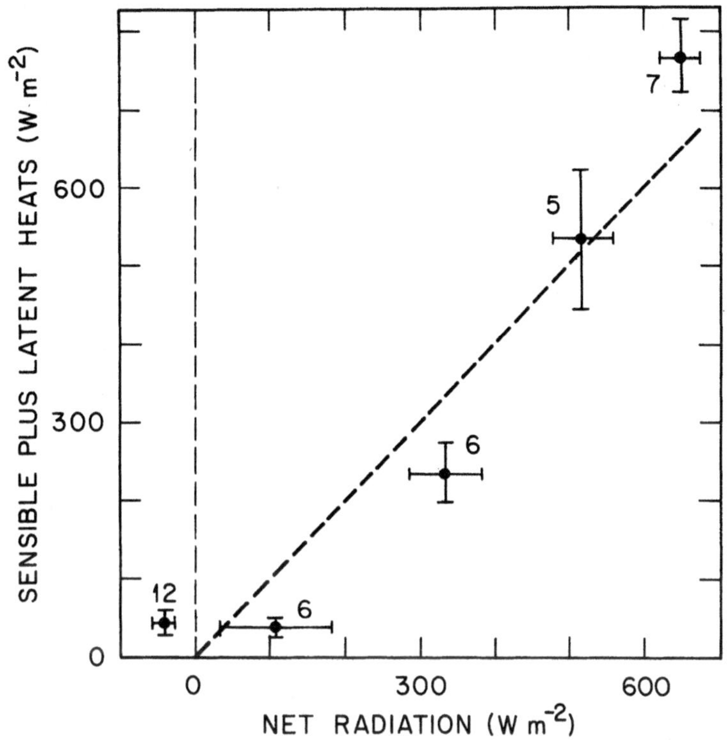

Fig. 3. Results of surface heat budget conducted at the NOAA/ARATDD forest field site in Oak Ridge, during May 1983. The line represents equality between the sum of sensible (H) and latent (LE) heat fluxes, and net radiation (R_n). Numbers indicate the size of the sample from which the plotted averages and standard errors are computed.

the data of Figure 3. The slope of the line is determined by theory, given independent measurement of the biomass per unit area of the canopy in question (41 kg m^{-2}) and an assumed specific heat of 1 cal g^{-1} $°C^{-1}$. The intercept remains undetermined because in reality neither canopy biomass temperature nor canopy air temperature alone is a completely adequate measure of the effective canopy temperature.

7. Simulation Results

The data quoted in Table I are typical of results obtained in daytime, convective conditions. The nocturnal case presents substantially greater difficulty, and has yet to be addressed in as much detail as the daytime case.

Table I also lists normalized covariances, which are directly related to the corresponding eddy fluxes, as derived from data smoothed by a range of time constants, as long as 30 s. Concentration standard deviations yielded by the same data-treatment procedures are also listed. The problems associated with the blind application of Equation (5) (or its equivalents), using standard deviations of turbulent fluctuations as

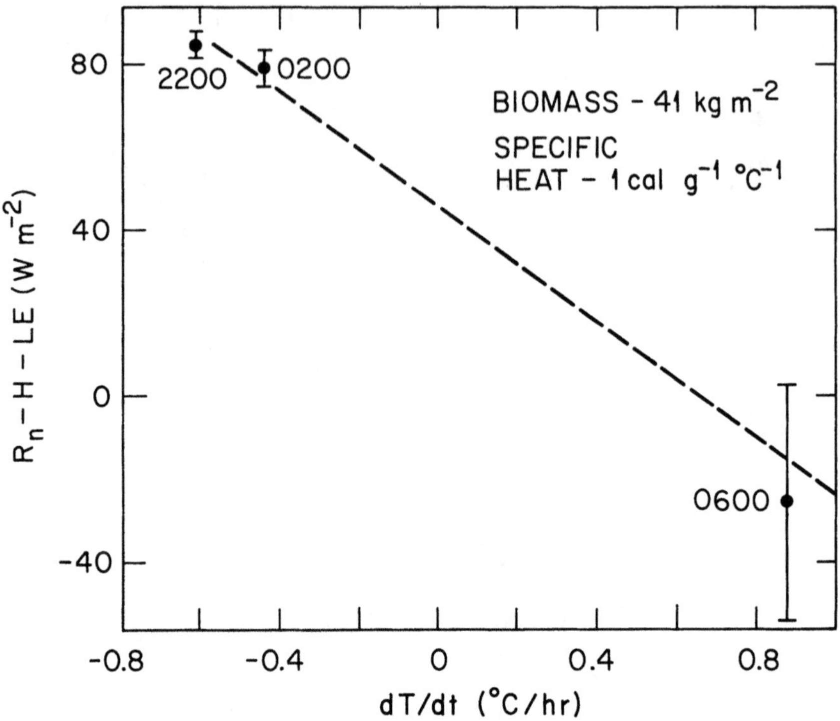

Fig. 4. The dependence of the surface heat flux residual ($R_n - H - LE$) or the time rate of change of canopy temperature. The line drawn corresponds to measured canopy biomass (41 kg m^{-2}) and an assumed specific heat of 1 cal g^{-1} °C^{-1}.

a basis for apportioning available heat energy and for evaluating pollutant fluxes, are evident from inspection of the values tabulated.

Figure 5 illustrates the consequences of high-frequency noise on the general utility of standard deviations as a basis for calculations. The values plotted are normalized geometric mean standard deviations for temperature (humidity values are indistinguishable), SO_2 and small particles (in the two bands used here, 0.5–0.7 μm and 0.7–1.2 μm diameter). As high-frequency noise is rejected by smoothing with increasing time constant, the effect on particle standard deviations is seen to be greatest. At the other extreme, the effect on temperature (and humidity) standard deviations is the least among the presently observed quantities. This is seen as a direct consequence of the ease with which temperature and humidity records can be obtained using modern sensors.

It should be noted that for particles, the problem is exacerbated by effects associated with poor counting statistics. The roll-off associated with the SO_2 data is less than that associated with the measurement of particle concentration, but greater than that for the meteorological variables. In this context, the use of covariances (with w) instead of standard deviations as indicators of total turbulent transport seems especially attractive, sicne the influence of high-frequency noise is then minimized.

Figure 6 extends the present considerations to covariances. The effect of sample response time on the evaluation of eddy fluxes is illustrated using SO_2 signals, and

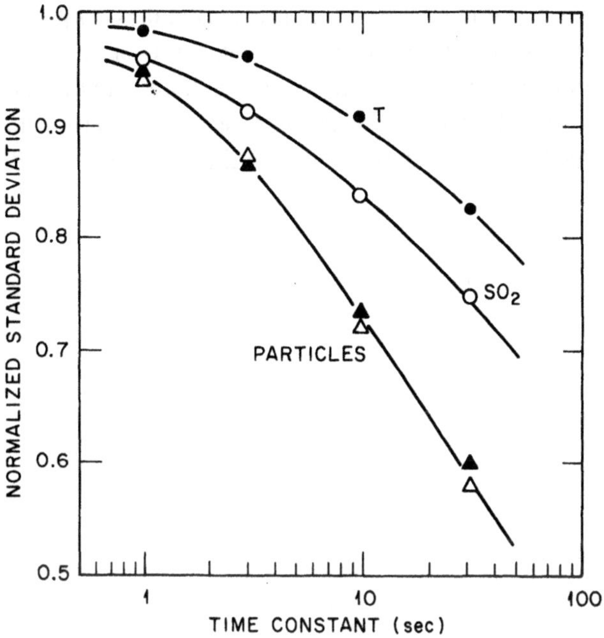

Fig. 5. The effect of sampling response time on apparent standard deviation associated with temperature (dots), sulfur dioxide (circles), and particle concentration (crosses) records. Data are normalized to values obtained from 'raw' data from fast-response sensors.

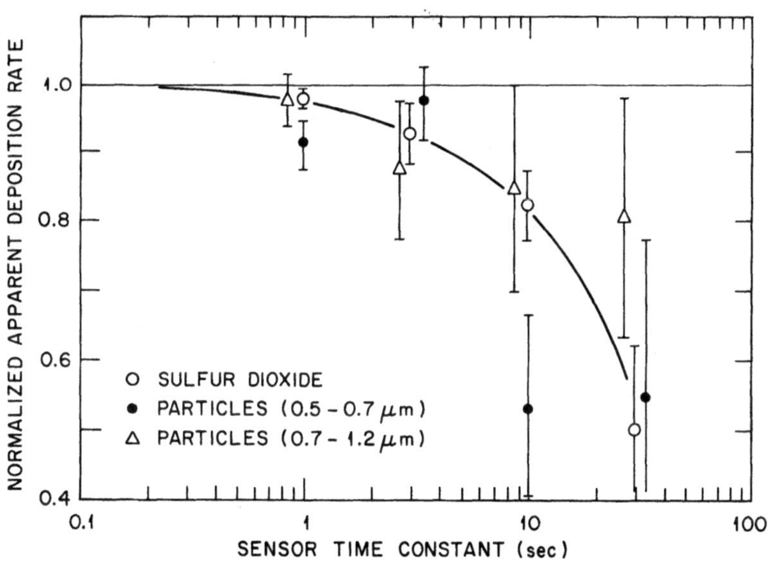

Fig. 6. The effect of sampling response time on the apparent vertical flux of SO_2 and small particles. Data plotted are normalized to values obtained from 'raw' data obtained using fast-response sensors.

particle records in the same size ranges as previously: 0.5–0.7 μm and 0.7–1.2 μm. As expected, covariances are less affected by inadequate sensor response than are standard deviations. It should be emphasized, however, that fluxes are the consequence of much faster fluctuations at night than in daytime, and hence the promise indicated by Figure 5 should not be extended beyond the daytime conditions in which the data were obtained.

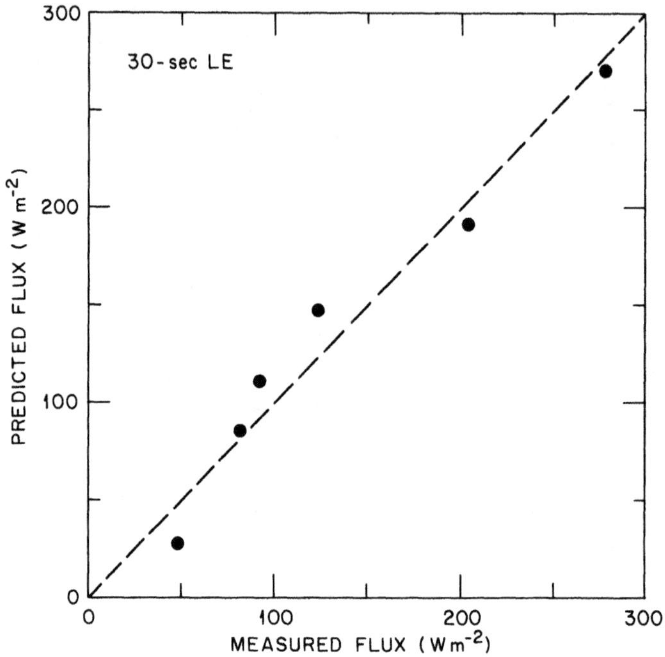

Fig. 7. A comparison of latent heat fluxes derived by application of (5), with direct-measurements made by eddy correlation. The heat storage term is assumed to be a constant 10% of the net radiation.

Figure 7 illustrates the ability of the similarity methods advocated here to provide accurate estimates of vertical fluxes. For purposes of illustration, Figure 7 plots estimated latent heat fluxes, *LE*, derived from using the most smoothed data (30-s time constant) listed in Table I, against the actual values measured by eddy correlation. In this case, the use of standard deviation data appears warranted, because of the relative ease with which a clean analog signal proportional to water vapor concentration can be produced, even at high frequencies (> 1 Hz). (For these studies, a Lyman-alpha hygrometer was employed.) Some scatter is evident in the diagram, but this scatter appears to be random and is much as expected on the basis of experience with comparisons between any pair of devices intended to measure the same turbulent flux.

8. Conclusions

The similarity methods discussed here are intended to permit fluxes of some trace species to be inferred from knowledge of other fluxes if appropriate gradient, turbulence intensity or covariance data are available to serve as a linkage. The constraint imposed by the surface heat energy balance involving (primarily) net radiation, sensible heat, latent heat of evaporation, and heat storage provides an opportunity to bypass the need for a direct flux measurement. In practice, net radiation is easily measured, using straightforward sensors available from commercial sources. No special micrometeorological skills are required for their proper use, although technical competence is certainly necessary. The method of present interest uses net radiation as a basis for calculations, and employs similarity principles first to derive the sensible heat flux and then to infer a dry deposition flux from knowledge of pollutant turbulence characteristics.

In concept, no fully-implemented micrometeorological turbulent flux data are required. Thus, the possibility exists (but remains to be fully explored) that the methods might be applicable in less ideal terrain than might otherwise be feasible.

Investigations of the methods described here are presently continuing; however, it is already apparent that the techniques summarized above are indeed capable of producing answers with reduced need for fully-expanded micrometeorological support, and without the usual eddy correlation sensor requirement for 1-s response or better. However, the analyses conducted so far point to some difficulties. If variances are used as the basis for computation, the effects of random noise propagate through the system in a manner that necessarily adds to the real signal variance, such that σ_c will always tend to be overestimated and, hence, so will v_d. This source of error is much the same as that which arises with other methods relying on single signal variances rather than on gradients or covariances.

It is also clear that the use of variance data as a basis for computations using the similarity/heat budget approach leads to uncertainty concerning the direction of the flux, precisely as in the case of direct analysis of variance in conventional meteorological applications.

An alternative method, in which the standard deviations in Equations (3)–(7) are replaced by covariances computed using some other, well-measured meteorological variable, appears capable of minimizing effects resulting from random noise, since by definition noise will be uncorrelated with any other signal. Furthermore, information on the sign of the vertical flux is then retained.

Acknowledgement

This work was conducted under the sponsorship of the U.S. Environmental Protection Agency, as a contribution to the National Acid Precipitation Assessment Program.

References

Baldocchi, D. D., Matt, D. R., McMillen, R. T., and Hutchison, B. A.: 1985, *Evapotranspiration from an Oak-Hickory Forest*, Proceedings, National Conference on Advances in Evapotranspiration, ASAE, St. Joseph, MI, pp. 414–422.

Dyer, A. J.: 1975, 'Measurement of Turbulent Fluxes by Fluxatron and NIFTI Techniques', *Atmos. Tech.* **7**, 24–29.

Dyer, A. J., and Hicks, B. B.: 1970, 'Flux-Gradient Relationships in the Constant Flux Layer', *Quart. J. Roy. Meteorol. Soc.* **96**, 715–721.

Garratt, J. R.: 1978, 'Flux Profile Relations Above Tall Vegetation', *Quart. J. Royal Meteorol. Soc.* **104**, 199–211.

Hales, J. M., Hicks, B. B., and Miller, J. M.: 1987, 'The Role of Research Measurement Networks as Contributors to Federal Assessment of Acid Deposition', *Bull. Amer. Meteorol. Soc.*, in press.

Hicks, B. B.: 1979, *Some Micrometeorological Methods for Measuring Dry Deposition Rates*, AICHE Symposium Series, 75, No. 188, 187–190.

Hicks, B. B.: 1985, 'Application of Forest Canopy – Atmospheric Exchange Information', in B. A. Hutchison and B. B. Hicks (eds.), The Forest-Atmosphere Interaction, D. Reidel Publ. Co., Boston, pp. 631–644.

Hicks, B. B., Hyson, P., and Moore, C. J.: 1975, 'A Study of Eddy Fluxes over a Forest', *J. Appl. Meteorol.* **14**, 58–66.

Hicks, B. B., Hess, G. D., and Wesely, M. L.: 1979, 'Analysis of Flux-Profile Relationships Above Tall Vegetation – An Alternative View', *Quart. J. Roy. Meteorol. Soc.* **105**, 1074–1077.

Hicks, B. B., Matt, D. R., McMillen, R. T., Womack, J. D., and Shetter, R. F.: 1983, in P. J. Samson (ed.), *Eddy Fluxes of Nitrogen Oxides to a Deciduous Forest in Complex Terrain*, Transactions, APCA Conference on Meteorology of Acidic Deposition, APCA, Pittsburgh, pp. 189–201.

Huebert, B. J. and Roberts, C. H.: 1985, 'The Dry Deposition of Nitric Acid to Grass', *J. Geophys. Res.* **90**, 2085–2090.

Kaimal, T. C., Wyngaard, T. C., Izumi, Y., and Cate, O. R.: 1972, 'Spectral Characteristics of Surface Layers Turbulence', *Quart. J. Roy. Meteorol. Soc.* **98**, 563–589.

Kanemasu, E. T., Wesely, M. L., Hicks, B. B., and Heilman, J. L.: 1979, 'Techniques for Calculating Energy and Mass Fluxes', in B. J. Barfield and J. F. Gerber (eds.), *Modification of the Aerial Environment of Crops*, American Soc. of Agric. Eng., pp. 156–182.

McCaughey, J. H.: 1985, 'Energy Balance Terms in a Mature Mixed Forest at Petawa, Ontario – A Case Study', *Boundary-Layer Meteorol.* **31**, 89–101.

Pruitt, W. O., Morgan, D. L., and Lourence, . J.: 1973, 'Momentum and Mass Transfer in the Surface Boundary Layer', *Quart. J. Roy. Meteorol. Soc.* **99**, 370–386.

Thom, A. S.: 1975, 'Momentum, Mass, and Heat Exchange of Plant Communities', in J. L. Monteith (ed.), *Vegetation and the Atmosphere*, Vol. I, Academic Press, London, pp. 57–109.

Wyngaard, J. C. and Coté, O. R.: 1972, 'Cospectral Similarity in the Atmospheric Surface Layer', *Quart. J. Roy. Meteorol. Soc.* **98**, 590–603.

THE STABLY STRATIFIED BOUNDARY LAYER OVER THE GREAT PLAINS

I. *Mean and Turbulence Structure*

DONALD H. LENSCHOW, XING SHENG LI[1], CUI JUAN ZHU[1], and B. BOBA STANKOV[2]

National Center for Atmospheric Research, P.O. Box 3000, Boulder, CO 80307, U.S.A.*

(Received in final form 29 August, 1987)

Abstract. We present airplane measurements of the stably stratified nocturnal boundary layer obtained during the Severe Environmental Storms and Mesoscale Experiment (SESAME) in 1979. The cases presented here were obtained over rolling terrain in central Oklahama, with a mean slope of about 0.003. The results are in general agreement with previous modeling and observational studies for the mean and turbulence structure of the nocturnal boundary layer, with the exception that the eddy diffusivity of heat, and consequently the flux Richardson number are less than expected.

1. Introduction

The mean and turbulence structure of the stably stratified boundary layer (herafter NBL – the nocturnal boundary layer) that exists over land at night in areas that are not overcast is still not very well understood compared to the daytime convective boundary layer. This is mainly due to suppression of vertical transport in a stably stratified fluid, which results in intermittent vertical exchange, particularly above the surface layer; and the relative importance of terrain effects on the mean flow. Yet, the NBL plays an important role in the development and evolution of nocturnal thunderstorms and squall lines, particularly during spring and early sumer over the lower Mississippi Valley. Furthermore, because of limited vertical diffusion, pollutants injected into a NBL are confined to a shallower layer and thus may be more concentrated than in the daytime. Other important characteristics of the NBL are the relatively large changes in wind speed and direction with height, the frequent occurrence of a wind maximum within the first kilometer that may be greater than the geostrophic wind at that level, and the rapid decrease of turbulence intensity with height.

Many of the gross features of NBL have been studied and documented (e.g., Mahrt *et al.*, 1979; Caughey *et al.*, 1979; Nieuwstadt, 1984). Some of these features have also been simulated by numerical models (e.g., Brost and Wyngaard, 1978). Both tethered balloons and towers have been used to study the mean and turbulence structure of a NBL (Caughey *et al.*, 1979; Nieuwstadt, 1984). Both have limitations. Neither is able to measure horizontal variations which are more significant in the NBL than the

[1] Present affiliation: National Marine Environmental Forecasting Center, P.O. Box 3303, Beijing, China.
[2] Present affiliation: Meteorological Wave Propagation Laboratory, NOAA, Boulder, CO 80303, U.S.A.
* The National Center for Atmospheric Research is sponsored by the National Science Foundation.

Boundary-Layer Meteorology **42** (1988) 95–121.

convective boundary layer, and existing towers are not tall enough to probe consistently through the top of nocturnal wind maxima.

With these limitations in mind, a NBL observational program was carried out as a subprogram of the Severe Environmental Storms and Mesoscale Experiment (SESAME) in May 1979 over central Oklahoma. The primary platform was an NCAR Queen Air aircraft, with additional measurements obtained from University of Virginia tethered balloon and acoustic sounder systems.

Part I deals with horizontally averaged measurements of boundary-layer variables, and includes analysis of both constant level flight legs as well as slant ascent and descent legs. For the most part, the observations are compared to previous observational studies and to modeling predictions. Part II discusses the horizontal variability and spectral aspects of the measurements. These are characteristics that are not available from tower or tethered balloon measurements and, therefore, comparisons with existing data sets are more limited.

2. Measurements

The aircraft flight tracks were oriented north-south mainly over the Fort Sill Military Reservation in central Oklahoma. The terrain is rolling hills a few tens of meters high covered mainly with grass interspersed with a few trees and buildings. Because of flight restrictions, the low-level flight legs were confined to hours with some daylight visibility within the military reservation. This limited the track length to typically about 20 km (five minutes flight time at $\sim 70 \text{ m s}^{-1}$). Seven flights were conducted. Early morning flights on 5, 6, 7, and 9 May and early evening flights on 4 and 5 May, 1979 have been analyzed. Mahrt (1985) discusses the early morning flight on 5 May, 1979, which is an example of a NBL, with very light mean horizontal winds. Here we report mainly on the early evening flight of 5 May and the early morning flights of 6 and 7 May. During this period, the region was under the influence of high pressure to the south and east, and the mean wind flow was southerly. An 850 mb analysis chart for 0600 LT on 6 May, 1979 is shown in Figure 1.

Measurements of the three air velocity components, temperature and humidity fluctuations at a sample rate of 20 s^{-1} are used in the analysis. Temperature was measured with a resistance wire thermometer corrected for dynamic heating, and humidity fluctuations were measured with a microwave refractometer. The instruments are described by Wyngaard *et al.* (1978). Lenschow (1986) discusses the methods used for measuring air velocity from an aircraft. The variables are filtered with 4-pole Butterworth analog filters with a cutoff frequency of 10 Hz to reduce the effects of aliasing. Static pressure, surface radiation temperature, and mean humidity from a dew-point hygrometer are available at a rate of 1 s^{-1}.

Both constant-level (above the ground) and slow ascent and descent flight legs were flown. One of the objectives of the ascent-descent flight legs was to obtain continuous vertical profiles of both mean and turbulence quantities throughout the stably stratified boundary layer. The long-wavelength maximum necessary to include the major contribu-

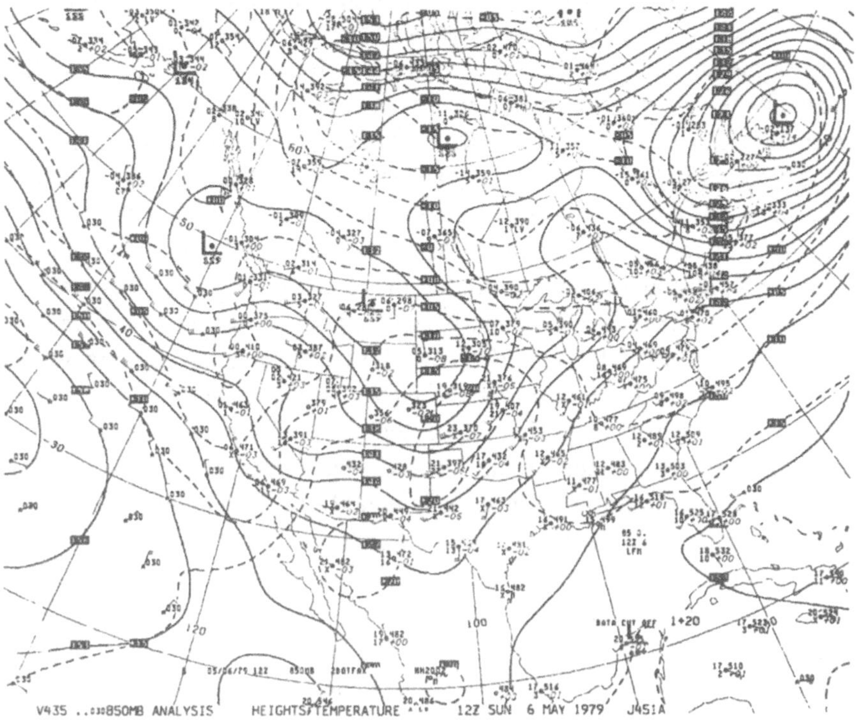

Fig. 1. Analysis chart for 850 mb at 0600 LT 6 May, 1979, during Flight 5.

tions to the fluxes and variances in the NBL is considerably less than in the unstably stratified boundary layer. Furthermore, since the height of the NBL is typically less than about 20% of the depth of the unstable boundary layer, fluxes and variances change more rapidly with height in the NBL. The changes with height are so rapid that it becomes difficult to resolve the height variation with constant-level legs at several levels in the boundary layer since the airplane height above the ground varies by as much as 30 m from both terrain changes and altitude variations due to turbulence and pilot-induced control changes; for these reasons, ascent-descent flight legs were used to resolve the vertical profiles of both mean and turbulence quantities.

On the other hand, the constant level legs provide a baseline for estimating spectral variables, as well as turbulence statistics, at particular levels in the boundary layer. Therefore, the analyses that follow utilize a combination of ascent-descent and constant level flight legs.

In addition to the turbulent fluctuations, ascent-descent profiles also contain mean variations due to the mean changes with height. The turbulent fluctuations are removed by high-pass filtering using a Lanczos filter (Lenschow and Stephens, 1980). The cutoff frequency of 0.1 Hz was selected to include the significant turbulent-scale variations, using for comparison the constant level legs. The average ascent-descent rate was

$\sim 2 \text{ m s}^{-1}$, which means that 0.1 Hz corresponds to a vertical displacement of $\sim$ 20 m. Over this height interval, the mean profiles can be considered well-approximated by a slowly varying function which is removed by the high-pass filtering except, perhaps, near the axis of the nocturnal jet or the inversion, which is usually in the vicinity of the nocturnal jet. The time series were then smoothed with a low-pass Lanczos filter with a cutoff frequency f_c of 0.01 Hz ($\sim$7 km horizontal distance and $\sim$200 m vertical displacement). Mahrt (1985) discusses errors introduced in both mean and turbulence statistics by estimating vertical structure from slant flight paths. This smoothing removes variations on individual soundings so that the results from several legs are more easily composited. A block diagram summarizing this data-processing procedure is shown in Figure 2. Other filter cutoffs were tried as well. Figure 3 compares results obtained with $f_c = 0.01$ and 0.02 Hz. We see that the higher frequency cutoff gives unsatisfactory results especially when computing the vertical temperature gradients and wind shear, because perturbations in the profiles caused zero values of the smoothed gradients which, in the case of wind shear, gives an infinite Richardson number. A more desirable way to obtain these measurements without the necessity for averaging for such long time periods is to use direct measurements of vertical temperature and wind differences, as discussed by Kristensen and Lenschow (1987).

In general, the results from constant-level legs (which averaged about 3 min or 13 km in length) were more scattered than the composited ascent-descent profiles. This may

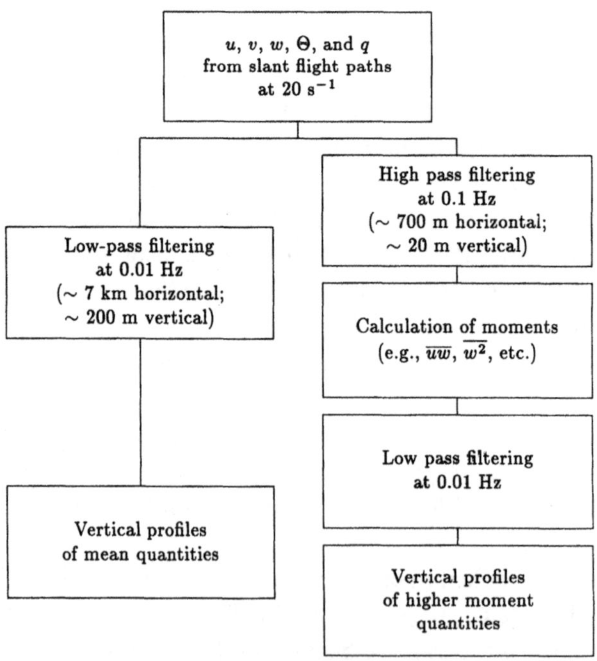

Fig. 2. Block diagram of the low- and high-pass filtering used for the 20 s^{-1} sample rate airplane data obtained from the slant ascent-descent flight paths.

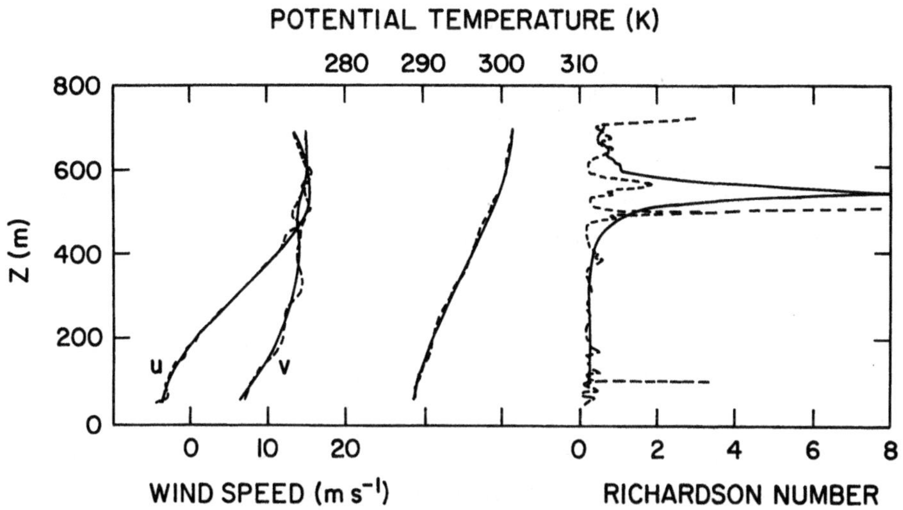

Fig. 3. Low-pass filtered wind velocity components and potential temperature, and Richardson number computed from the vertical gradients of the filtered wind components and potential temperature. The solid line was obtained using a cutoff frequency of 0.01 Hz, which was used for the results reported here; the dashed line was obtained using a cutoff frequency of 0.02 Hz.

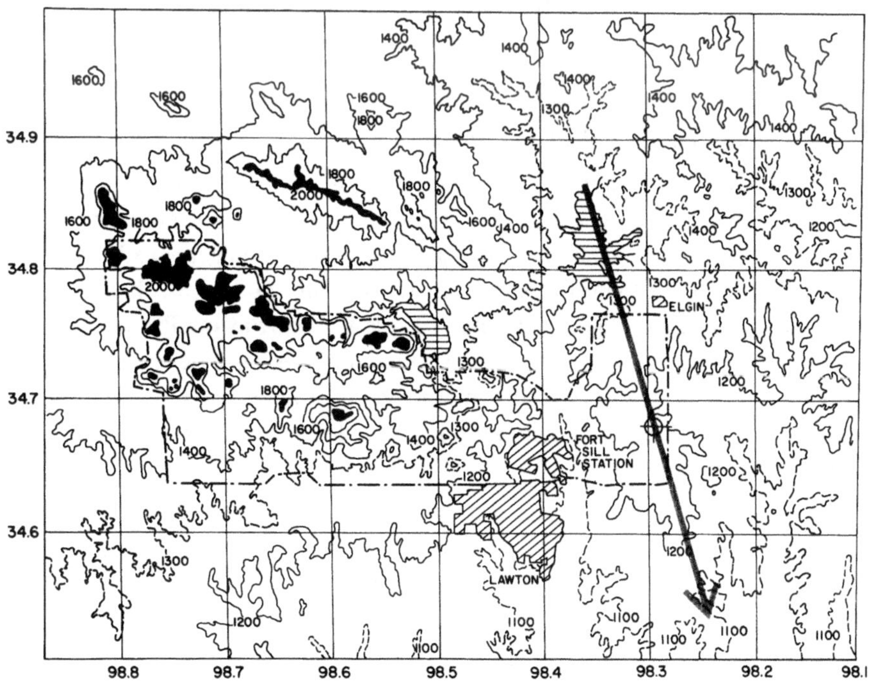

Fig. 4. Contour map of the experimental area. The contour interval is 200 feet (60 m) with supplementary dashed 100 feet (30 m) contour lines below 1400 feet (427 m). The arrow points in the direction of the terrain slope. The circle at the arrow's center is the location of the ground observation site, and is approximately at the center of the research flight area.

be, as discussed earlier, due to difficulties in maintaining a constant height above the surface. Both mean and turbulent quantities show considerable variation over the estimated 30 m height variation, which is $\simeq$ 10 to 20% of the total NBL height.

An additional complication is horizontal variation in the mean boundary-layer structure, which is at least partially related to variations in terrain height and land usage. One of the characteristics of the NBL, as compared to the convective boundary layer is the relatively greater importance of terrain variability on boundary-layer structure. As Brost and Wyngaard (1978) have shown, even with terrain slopes as small as 0.001, the boundary-layer structure is significantly modified. Figure 4 shows a contour map of the experimental area. The terrain slope on a scale of ~30 km is estimated to be about 0.003 at an azimuth angle of 160°. It is difficult to decide over what scale to calculate the terrain slope that is most relevant for estimating terrain slope effects. Caughey *et al.* (1979) selected a region of 13 km by 16 km for estimating the slope for calculating terrain effects on the Minnesota Experiment. Here we have chosen a scale for estimating the effects of mean terrain slope on boundary-layer structure that corresponds to an advection time of about one hour; the slope over a larger area would not be very much different. We also investigate the effects of baroclinity and different nocturnal cooling rates on the boundary-layer structure.

3. Vertical Profiles of Second-Order Moments

3.1. Flat Terrain

3.1.1. *Turbulence Fluxes of Heat and Momentum*

In order to study in more detail the vertical structure of turbulence quantities, we first present a somewhat more generalized version of the analytical model developed by Nieuwstadt (1984, 1985), which assumes a horizontally homogeneous boundary layer with no terrain slope. We then compare predictions of this model with observations from the constant-level legs for Flights 5, 6, and 7. Table I shows the four flights used for most of the subsequent analysis and some of the boundary-layer parameters for these flights. Most previous measurements of NBL structure were obtained from fixed-point observations (e.g., towers and tethered balloons). In the following discussion, we show that in

TABLE I

List of airplane flights and boundary-layer parameters obtained from the airplane measurements

Flt. No.	Date (LT)	u_* ($m s^{-1}$)	L (m)	h (m)	d	$\overline{w\theta}_0$ ($K m s^{-1}$)
4	1733–1945 5 May	0.18	63	200	0.54	−0.0079
5	0530–0757 6 May	0.33	366	300	0.25	−0.0071
6	0518–0735 7 May	0.44	447	400	0.26	−0.016
7	0535–0724 9 May	0.38	428	600	0.43	−0.011

some cases, the aircraft measurements support fixed-point observations; at the same time, the mobility of the aircraft allows us to measure horizontal variability, which is unavailable from the fixed-point observations. These results are presented in Part II. Most of the subsequent analysis will deal with Flights 4, 5, and 6; averaged profiles for wind speed and direction, potential temperature, humidity, Richardson number, and velocity and temperature variances for these flights obtained from the airplane slant flight paths, are presented in Figure 5.

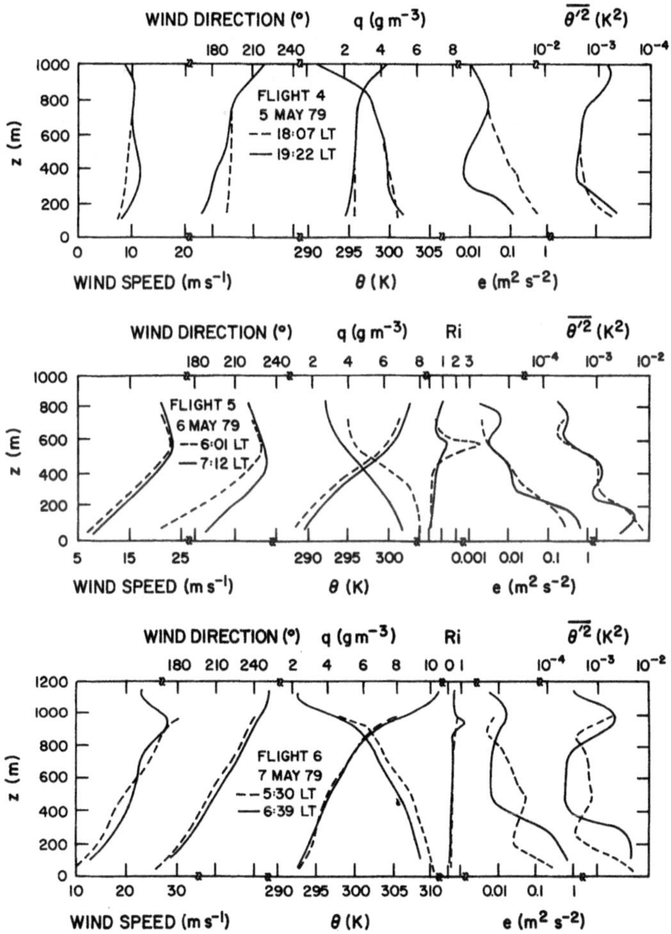

Fig. 5. Filtered vertical profiles of wind speed and direction, potential temperature, Richardson number (for Flights 5 and 6), turbulence energy and temperature variance for Flights 4, 5, and 6, obtained from the airplane slant flight paths.

The vertical profile of the temperature flux over flat terrain can be obtained from the thermal energy equation:

$$\frac{\partial \theta}{\partial t} = -\frac{\partial \overline{w\theta}}{\partial z} + \left(\frac{\partial \theta}{\partial t}\right)_r, \qquad (1)$$

where $(\partial\theta/\partial t)_r$ is cooling due to radiation. Just before sunrise, the airplane observations indicate that $|\partial\theta/\partial t| < 0.4$ C hr^{-1}. Furthermore, Figure 11 (Part II) indicates that the air temperature increases by less than a degree in the first hour after sunrise. Therefore, we assume steady-state conditions, so that (1) becomes

$$-\frac{\partial\overline{w\theta}}{\partial z} + \left(\frac{\partial\theta}{\partial t}\right)_r = 0 . \tag{2}$$

Radiational cooling is a complicated function of the temperature and humidity structure through the boundary layer and above. Previous analytical models of the NBL have either neglected radiational cooling (e.g., Nieuwstadt, 1984) or assumed a linear decrease of radiational cooling to zero at the top of the NBL (e.g., Yamada, 1979; Nieuwstadt, 1980). We assume here a somewhat more general expression for radiational cooling. We start by noting that the air very near the surface is cooled by a combination of radiational cooling and convection. Therefore,

$$\left(\frac{\partial\theta}{\partial t}\right)_r (0) = C\,\frac{\partial\theta_0}{\partial t} , \tag{3}$$

where $\partial\theta_0/\partial t$ is the total (radiative plus convective) cooling rate near the surface and C is a constant between 0 and 1.

Since turbulence in the NBL is assumed to be negligible at the top of the boundary layer $z = h$, we have at this height

$$\left(\frac{\partial\theta}{\partial t}\right)_r = \frac{\partial\theta_h}{\partial t} . \tag{4}$$

A typical value for $\partial\theta_h/\partial t$ is about 2 C day^{-1} (Yamada, 1979). This is about an order of magnitude smaller than $\partial\theta_0/\partial t$. Therefore, $\partial\theta_h/\partial t$ can be neglected compared to $\partial\theta_0/\partial t$. We then assume the more general expression for radiational cooling,

$$\left(\frac{\partial\theta}{\partial t}\right)_r = C\,\frac{\partial\theta_0}{\partial t}\,(1 - z/h)^\nu , \tag{5}$$

where ν is a constant. This is in general agreement with the radiational cooling rates obtained by André *et al.* (1978) and Garratt and Brost (1981) from numerical models. This type of radiational cooling profile is characteristic of a boundary layer without a strong capping inversion, with constant or decreasing humidity with height, and with decreasing curvature in the lapse rate with height. This seems not inconsistent with the structure observed here.

We introduce the dimensionless variables:

$$H = \overline{w\theta}/(\overline{w\theta})_0 , \tag{6}$$

$$\eta = z/h , \tag{7}$$

$$\sigma_1 = \frac{\partial\theta_0}{\partial t}\,\frac{h}{(\overline{w\theta})_0} , \tag{8}$$

where $(\overline{w\theta})_0$ is the surface temperature flux. Rewriting the thermal energy equation (1) in terms of the normalized variables, we have

$$\frac{\mathrm{d}H}{\mathrm{d}\eta} = C\sigma_1(1-\eta)^{\nu}. \tag{9}$$

The boundary conditions are

$$H = 1 \quad \text{for} \quad \eta = 0, \qquad H = 0 \quad \text{for} \quad \eta = 1. \tag{10}$$

A solution to (9) which satisfies the boundary conditions (10) is

$$H = (1-\eta)^r, \tag{11}$$

where $r = \nu + 1$. Nieuwstadt (1984) found that for his observations, $r = 1$. This is somewhat surprising in view of the role that radiational cooling plays in the NBL (e.g., André and Mahrt, 1982; Garratt and Brost, 1981). Radiational cooling should result in a temperature flux profile that has an increasing slope with height (i.e., a positive curvature). Our observations, based on the constant-level aircraft legs, give a value of about $r = 1.5$, as shown in Figure 6. The results of Caughey *et al.* (1979) are consistent with an even larger value of r.

We can carry out a similar analysis for the momentum flux profile, again following Nieuwstadt (1984), but introducing r as a parameter that can vary, depending upon the relative magnitude of radiational cooling. We assume, as did Nieuwstadt (1984), that

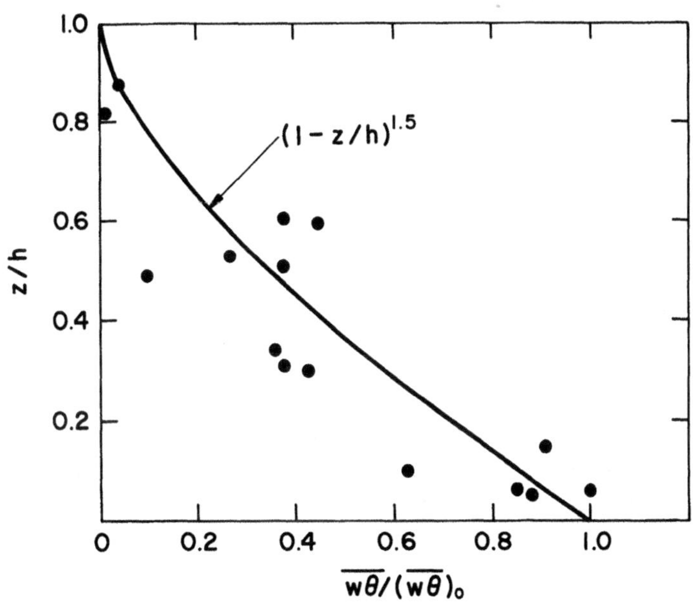

Fig. 6. Turbulence heat flux normalized by the values at the lowest flight levels for Flights 4–7.

the flux Richardson number is constant; i.e.,

$$\mathrm{Ri}_f = \frac{(g/T)\overline{w\theta}}{r\Omega} = \text{constant}\,, \tag{12}$$

where

$$r\Omega = -\overline{uw}\,\frac{\partial U}{\partial z} - \overline{vw}\,\frac{\partial V}{\partial z}\,. \tag{13}$$

In fact, as we discuss later, Ri_f is not observed to be constant here, but increases somewhat with height. We introduce the additional dimensionless variables (Nieuwstadt, 1984)

$$\sigma = -\overline{uw}/u_*^2 + i(-\overline{vw}/u_*^2)\,, \tag{14}$$

$$S = \frac{L}{u_*}\left(\frac{\partial U}{\partial z} + i\,\frac{\partial V}{\partial z}\right), \tag{15}$$

where $L \equiv -u_*^3/[(g/T_0)k(\overline{w\theta})_0]$ is the Monin–Obukhov length. Introduction of these normalized variables into the equations

$$f(V - V_g) - \frac{\partial\overline{uw}}{\partial z} = 0\,, \tag{16}$$

$$-f(U - U_g) - \frac{\partial\overline{vw}}{\partial z} = 0\,, \tag{17}$$

and the assumption that the geostrophic wind is constant with height lead to

$$\frac{\mathrm{d}^2\sigma}{\mathrm{d}\eta^2} - id^2S = 0\,, \tag{18}$$

$$\frac{H}{\sigma^* S} = k\,\mathrm{Ri}_f, \tag{19}$$

where σ^* is the complex conjugate of σ and

$$d = h(u_* L/f)^{-1/2}\,. \tag{20}$$

As shown in Figure 13, the assumption of constant geostrophic wind is not really valid in this region primarily because of the terrain slope. We also, for mathematical convenience, neglect the drainage flow terms contributed by the terrain slope which are given in (35) and (36), and which also act like a height-varying geostrophic wind. The boundary conditions are

$$\sigma = 1 \quad \text{for} \quad \eta = 0\,, \qquad \sigma = 0 \quad \text{for} \quad \eta = 1\,. \tag{21}$$

Combining (18) and (19) with (11), we have

$$\sigma^* \frac{\mathrm{d}^2\sigma}{\mathrm{d}\eta^2} - i \frac{\mathrm{d}^2}{k\,\mathrm{Ri}_f} (1 - \eta)^r = 0\,. \tag{22}$$

We assume a solution of the form

$$\sigma = (1 - v)^{a + ib}\,. \tag{23}$$

Substituting (23) into (22), we find that

$$a = \tfrac{1}{2}(r + 2)\,, \tag{24}$$

$$b = \frac{r_1}{r + 1} = \sqrt{a(a - 1)}\,, \tag{25}$$

$$d^2 = (1 + r)\,[r(2 + r)]^{1/2} k\,\mathrm{Ri}_f/2\,, \tag{26}$$

where

$$r_1 = d^2/(k\,\mathrm{Ri}_f)\,. \tag{27}$$

Thus, the magnitude of the normalized vertical profile of momentum flux r/u_*^2, which satisfies the boundary condition (24) is given by

$$r/u_*^2 = (1 - \eta)^a = (1 - z/h)^{1.75}\,. \tag{28}$$

We see in Figure 7 that (28) is consistent with the observations; albeit, the considerable

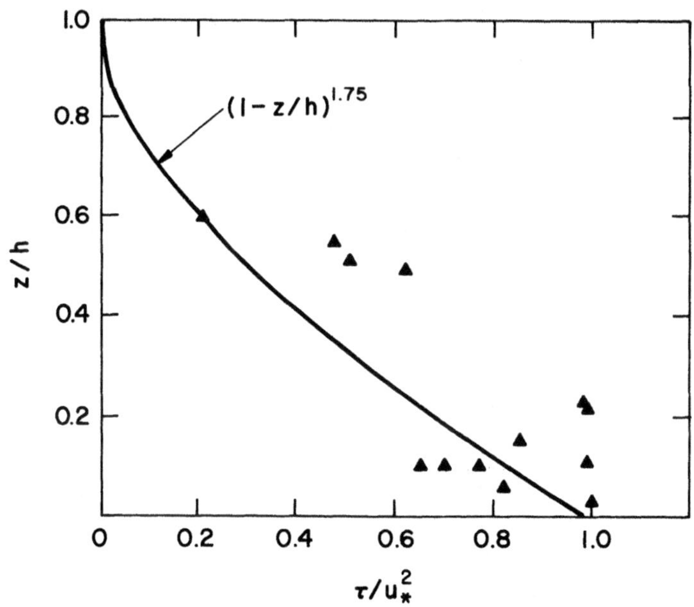

Fig. 7. Momentum flux normalized by the values at the lowest flight levels for Flights 4–7.

scatter means that the results are consistent with a wide range of exponents, and with the observations of Caughey *et al.* (1979), as well as Nieuwstadt (1984), who obtained the value $a = 1.5$ by assuming a linear decrease in heat flux with height. As noted by Lenschow and Stankov (1986), momentum flux estimates in an unstably stratified boundary layer have considerably more scatter than scalar fluxes such as temperature; this is also likely to be true in the NBL.

One significant difference between the results reported here and those of Nieuwstadt (1984) is that $\mathrm{Ri}_f \simeq 0.08$ through most of the boundary layer, which is considerably less than the limiting value of 0.2 that Nieuwstadt obtained. Values of Ri_f were computed from the ratios of the smoothed temperature flux to the sum of the products of the two smoothed momentum flux components times the wind shear component evaluated from the smoothed mean wind component profiles at 50 m height intervals. Other investigators (e.g., Kondo *et al.*, 1978; Ueda *et al.*, 1981) have also obtained values of Ri_f considerably less than 0.2 in stable conditions. However, as noted in Part II, terrain effects may play an important role in the structure of the boundary layer observed here. It seems plausible that the observed terrain-related wave effects may be relatively more important for momentum transport than for heat transport. We also cannot exclude the fact that Ri_f may be sensitive to the filter cutoff, although we did not note any systematic effect of filter cutoff.

The parameter d is the constant of proportionality for specifying the boundary-layer height as first proposed by Zilitinkevich (1972), i.e.:

$$\frac{hf}{u_*} = d\left(\frac{u_*}{fL}\right)^{-1/2} . \tag{29}$$

Figure 8 is a plot of d versus r. For $r = 1$ and $k = 0.35$ as Nieuwstadt (1984) assumed, and using the value of $\mathrm{Ri}_f = 0.08$, we have $d = 0.22$. For $r = 1.5$, as we suggest here, $d = 0.28$, which is in reasonable agreement with the observed values given in Table I, especially for Flights 5 and 6, for which we have the most complete data.

The question of how to define the boundary-layer height h in a NBL is much more difficult in practice than in the convective boundary layer. This is illustrated by the different ways in which it has been defined in previous studies. Caughey *et al.* (1979), for example, use the height at which the temperature flux falls to 5% of its surface value, while Nieuwstadt (1984) uses the height of the echo region measured by an acoustic sounder.

We found it very difficult to estimate the boundary-layer height from the acoustic sounder records obtained during this experiment because of the lack of a well-defined top to the echo region. Instead, the signal intensity decreases gradually with height, and intermittent echo regions become evident. During Flight 4, the echo region almost disappears around 1900 LT, which marks the end of the day's convective activity, and slowly develops after about 1930 LT, but does not exceed about 60 m depth until after 2030 LT. During Flight 5, a strong echo region extended up to about 120 m, and above this, layers of about 50 m depth were present at about 160 and 320 m, and a faint layer

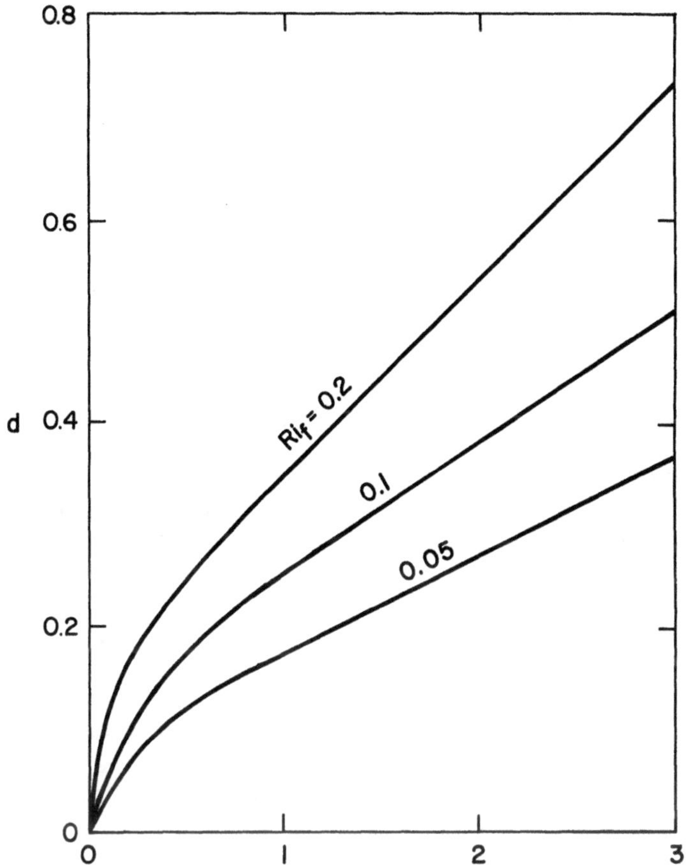

Fig. 8. Scaling constant for the nocturnal boundary-layer height d as a function of the temperature flux profile exponent r for several values of Ri_f.

at about 430 m. These layers vary considerably in height and intensity throughout the flight period, sometimes even merging with the surface-based echo region. Convection is evident after about 0800 LT. On the night preceding Flight 6, the shape of the echo region resembled a gaussian curve, starting out near the surface early in the evening, extending beyond 500 m (the maximum range of the sounder) in the middle of the night, then decreasing to between 100 and 200 m around 0600 LT. Convective activity is visible again after about 0800 LT.

Because of the ambiguity of the acoustic sounder record, and the scatter in the temperature flux measurements, especially at levels where the flux is small, we chose to use the point at which the filtered velocity variance $e = \frac{1}{2}(\overline{u^2} + \overline{v^2} + \overline{w^2})$ is 5% of its value near the surface as a measure of boundary-layer height.

After the evening transition, the Brost and Wyngaard (1978) model predicts that d tends to decrease with time from about 0.6 2 hours after the transition to about 0.4 5 hours after the transition, as shown in Figure 9. Our results (Table I and Figure 9) are consistent with this picture, since the three early-morning cases are closer to the 5-hour

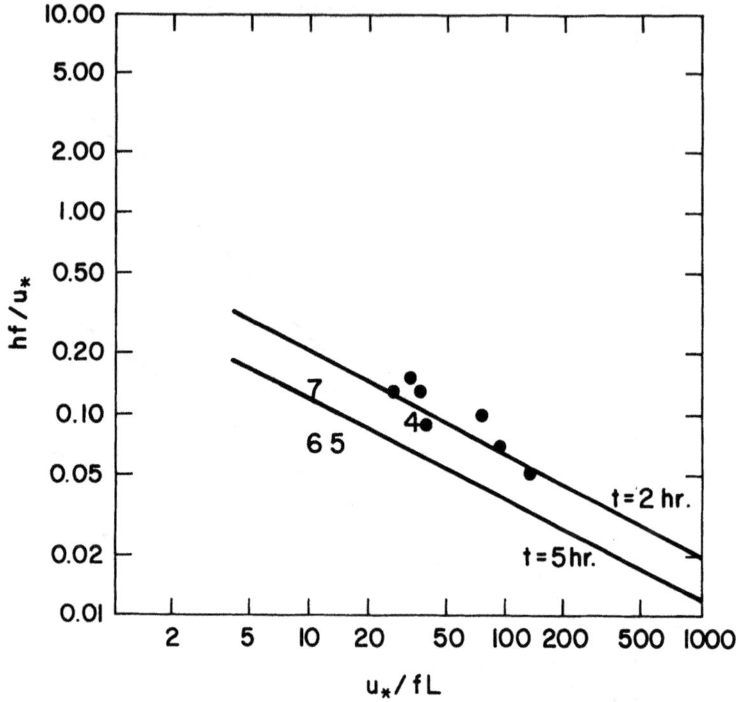

Fig. 9. Normalized boundary-layer height for Flights 4–7, denoted by the numbers. The results of Caughey *et al.* (1979) are denoted by the filled-in circles. The straight lines are predictions of the Brost–Wyngaard (1978) model for a surface cooling rate of 4 K h^{-1} at 2 hr and 5 hr after after transition.

curve, while the evening case is closer to the 2-hour curve. The average value of d for all four cases is 0.37 which is consistent with the observations presented by Garratt (1982) and Caughey *et al.* (1979), although there is considerable scatter in the observations.

3.1.2. *Local Similarity*

Following Nieuwstadt (1984), Monin–Obukhov similarity theory can be extended to the entire NBL by using local turbulence scales $L(z)$, $\tau(z)$, and $\theta_*(z)$ in place of surface-layer scales. All similarity functions are then assumed to have constant values; i.e., $\sigma_i^2(z)/\tau(z) = C_i$, $e(z)/\tau(z) = C_e$ and $\overline{\theta^2(z)}/\theta_*^2(z) = C_\theta$, where i signifies any of the three velocity components. With (11) and (28), this leads to

$$\sigma_i^2/u_*^2 = C_i(1 - z/h)^a = C_i(1 - z/h)^{1.75}, \tag{30}$$

and thus

$$e(z)/u_*^2 = C_e(1 - z/h)^{1.75}. \tag{31}$$

From the results presented in Figure 10 for Flights 5, 6, and 7, we find that $C_u = C_v = 4.5$, $C_w = 3.1$, and $C_e = 6$. Nieuwstadt (1984) found that for these observations, $a = 1.5$, $C_u = 4.2$, $C_v = 2.9$, $C_w = 2.0$, and $C_e = 4.6$, while Pasquill and Smith

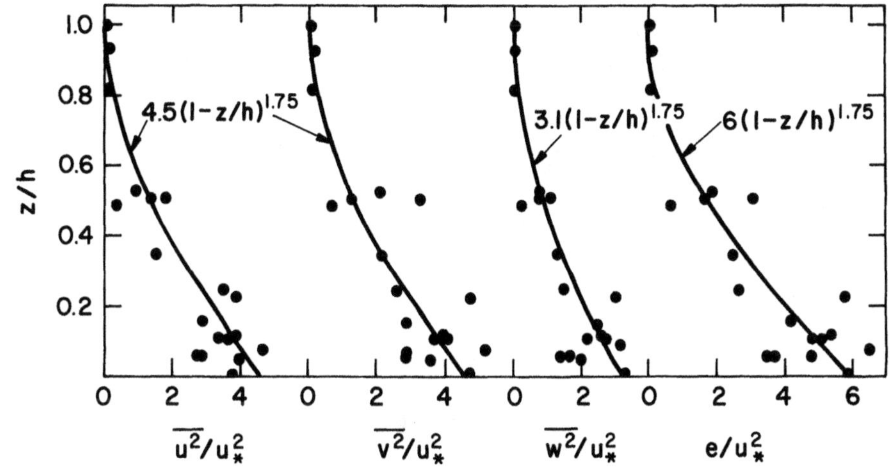

Fig. 10. Normalized velocity variances for Flights 4–7.

(1983) estimated values for the observations presented by Caughey *et al.* (1979) of $a = 1$, $C_u = 3.8$, and $C_w = 1.7$. Our results indicate that $a > 1$, but we cannot distinguish between $a = 1.75$ and $a = 1.5$. Lacser *et al.* (1986) obtained the values $a = 1.32$ and $C_e = 4.92$ from modeling results, which agrees reasonably well with ours and Nieuwstadt's observations.

Similarly, the local scaling hypothesis for the standard deviation of temperature gives $-\sigma_\theta \tau^{1/2}/\overline{w\theta} = C_\theta$, which leads to

$$\sigma_\theta/T_*(1 - z/h)^{r - a/2} = C_\theta(1 - z/h)^{5/8}, \tag{32}$$

where the temperature scale $T_* \equiv -\overline{w\theta}/u_*$ and C_θ is a constant. For the data presented in Figure 11, $C_\theta = 2$, with more scatter than for the velocity variance.

Again, following the analysis of Nieuwstadt (1985), we can calculate the normalized eddy diffusivity coefficients for momentum and temperature. We assume, as did Nieuwstadt, that Ri is constant with height and $\simeq 0.2$. Figure 5 shows that Ri increases with height from a value close to zero near the surface to about 0.2 in the upper half of the boundary layer. This is consistent with Nieuwstadt's (1985) observational results. The eddy diffusivity coefficients turn out to be independent of r, and are identical to Nieuwstadt's expressions,

$$K_m/(u_* L) = k\,\mathrm{Ri}_f(1 - z/h)^2, \tag{33}$$

$$K_h/(u_* L) = k\,\frac{\mathrm{Ri}_f^2}{\mathrm{Ri}}\,(1 - z/h)^2. \tag{34}$$

We note that as a consequence of this procedure for obtaining the diffusivities, we (as did Nieuwstadt, 1985) obtain an infinite temperature gradient at $z = h$, as pointed out by Sorbjan (1987). Therefore, these expressions do not apply near the top of the NBL. Values of $K_m/(u_* L)$ and K_h/K_m are plotted in Figure 12 from the slant flight path data

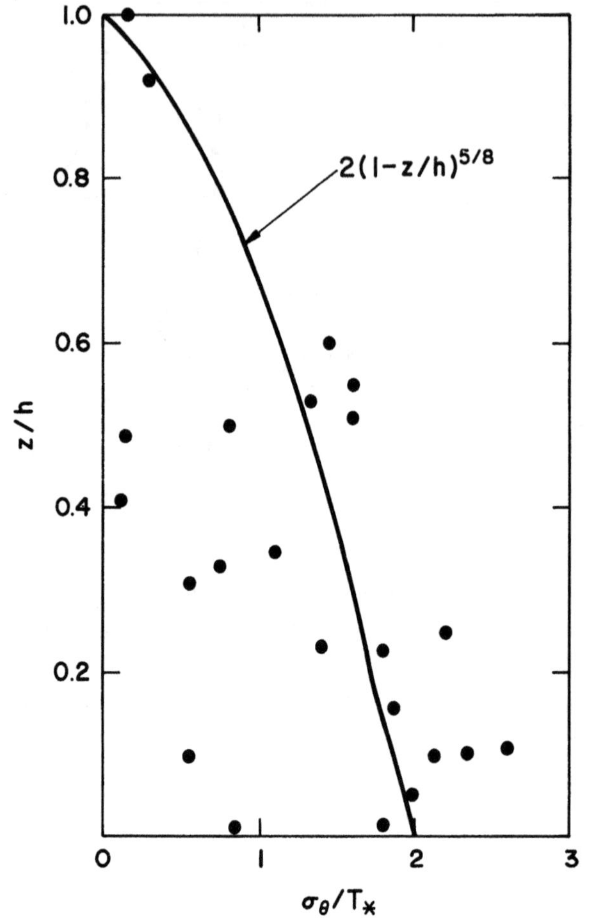

Fig. 11. Normalized temperature variances for Flights 4–7.

for Flights 5 and 6. The values $K_m/(u_*L)$ agree reasonably well with (33), and the ratio K_m/K_h is consistent with $\mathrm{Ri}_f \simeq 0.08$.

3.2. Large-scale terrain slope

As has been pointed out by many previous investigators (e.g., Bonner *et al.*, 1968), the nocturnal low-level jet can be considerably faster than geostrophic. Figure 13, which was obtained from the SESAME rawindsonde network data, shows that this is the case for Flights 5 and 6. In the Great Plains region, this effect is accentuated by the terrain slope. Paegle and Resch (1973) modeled this situation, including the effects of nonlinear terms arising from the horizontal variation of the horizontal pressure gradient and found that the wind can easily be more than 50% greater than geostrophic. Their model predicts the strongest southerly jet in the western portions of anticyclonic circulations. In Flight 5, the nocturnal jet is located at that portion of the anticyclonic circulation

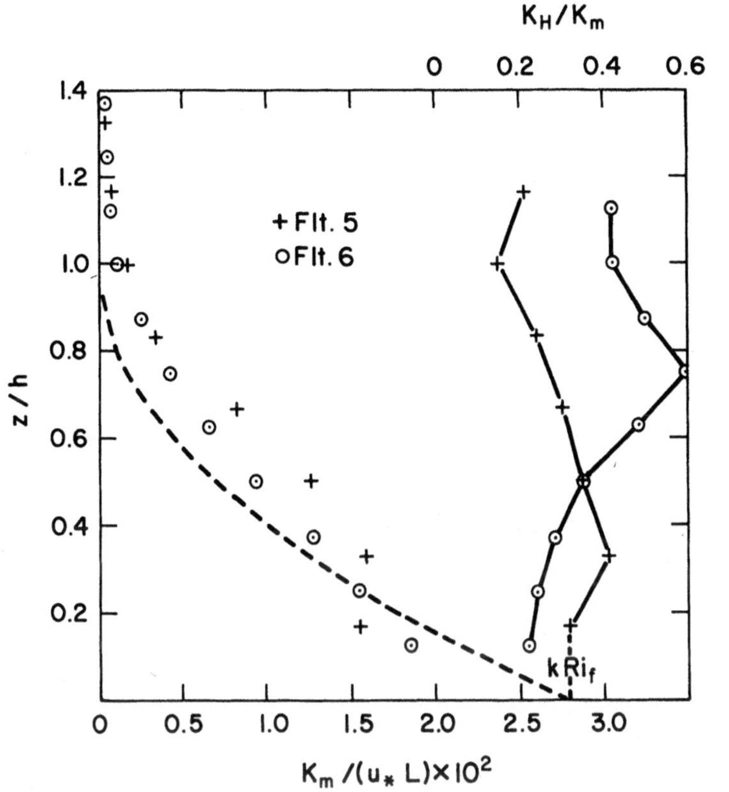

Fig. 12. Normalized eddy diffusivity for momentum and the ratio of diffusivities for heat to that for momentum for Flights 5 and 6. The dashed line is the formulation predicted by Nieuwstadt (1985).

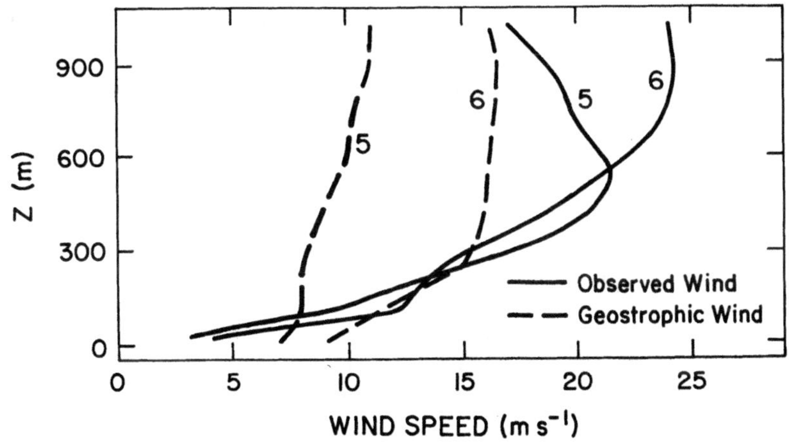

Fig. 13. Observed mean wind speed profile and geostrophic wind for Flights 5 and 6.

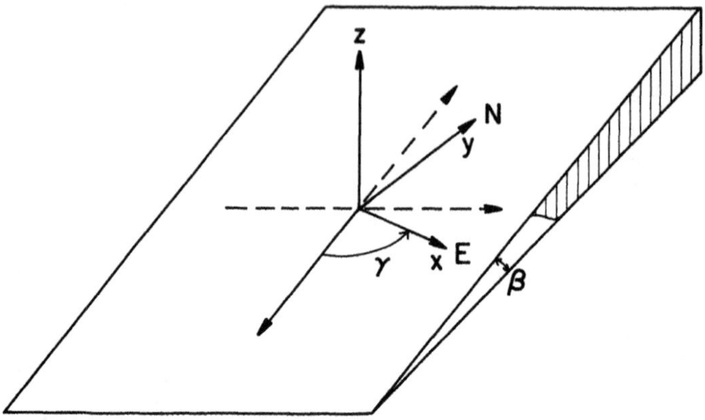

Fig. 14. Schematic of axes and angles for defining flow variables over sloped terrain.

(Figure 1) where both diurnal oscillations of eddy stress and buoyancy forces associated with sloping terrain contribute to its supergeostrophic speed.

Figure 14 illustrates the coordinate system used here for sloped terrain. The x- and y-axes are in the plane of the surface, with the x-axis parallel to the mean wind at the surface, and z is upwards, perpendicular to the surface. The slope angle is β, and the angle (measured counterclockwise) between the fall line vector and the x-axis is γ. The equations of motion can then be written as (Brost and Wyngaard, 1978)

$$\frac{\partial U}{\partial t} = -fV_g + fV - \frac{\partial \overline{wu}}{\partial z} - \frac{g}{T_0}\,\overline{T}'\,|\beta|\cos\gamma - U_i\,\frac{\partial U}{\partial x_i}\,, \tag{35}$$

$$\frac{\partial V}{\partial t} = fU_g - fU - \frac{\partial \overline{wv}}{\partial z} + \frac{g}{T_0}\,\overline{T}'\,|\beta|\sin\gamma - V_i\,\frac{\partial V}{\partial x_i}\,, \tag{36}$$

where the prime denotes a departure from the reference adiabatic state, T_0 is the reference temperature and the subscript i denotes a summation over all three components. We can use these equations to calculate the rate of change of wind direction (Neumann, 1977; Caughey *et al.*, 1979). Defining $\delta = \tan^{-1}(V/U)$ (i.e., 90° − wind direction, which means that $\partial\delta/\partial t > 0$ implies decreasing wind direction angle and *vice versa*), we have

$$\frac{\partial \delta}{\partial t} = S^2\left(U\,\frac{\partial V}{\partial t} - V\,\frac{\partial U}{\partial t}\right), \tag{37}$$

where $S^2 = U^2 + V^2$. Substituting (35) and (36) into (37), we have

$$\frac{\partial \delta}{\partial t} = S^{-2}\left(V\,\frac{\partial \overline{wu}}{\partial z} - U\,\frac{\partial \overline{wv}}{\partial z}\right) - U_i\,\frac{\partial \delta}{\partial x_i} + f[S^{-2}(UU_g + VV_g) - 1] +$$

$$+ \frac{g}{T_0}\,\overline{T}'S^{-2}\,|\beta|\,(U\sin\gamma + V\cos\gamma)\,. \tag{38}$$

The first term on the right-hand side is the turbulence friction term, the second is the advection term, the third is the pressure-coriolis term and the last is the drainage flow term. Caughey *et al.* (1979) have evaluated this expression in the surface layer using tethered balloon and tower measurements, but were not able to estimate the advective term. Here we use both the aircraft and SESAME rawindsonde data to estimate all the terms in (38) for Flights 5 and 6, except for the vertical component of the advection. The horizontal flight legs of the aircraft, which were in a N–S direction, were used to estimate the horizontal advection. This means that the estimates near the surface, where the wind direction is close to 180°, are more accurate than at higher levels, where the wind is more westerly. Although we did not estimate the vertical advection term, we note that in order for this term to be significant compared to the other terms, the mean vertical velocity would have to be $> 5 \times 10^{-3}$ m s^{-1}, which is not likely to be the case.

Figure 15 shows that all terms contribute significantly to the angle change throughout the boundary layer. For Flight 5, the wind direction varied with height from SSW to

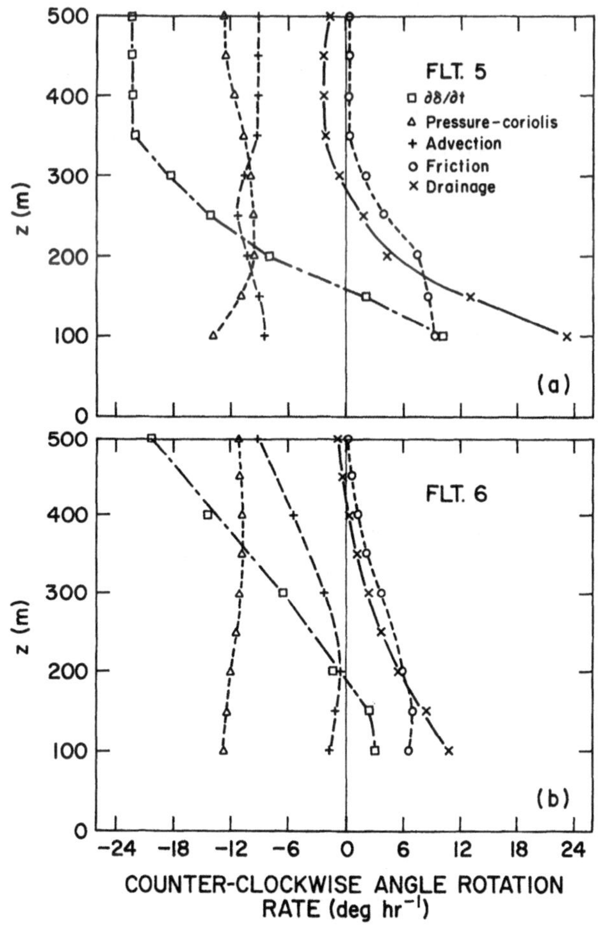

Fig. 15. Terms in the equation for rotation of the horizontal wind vector. Panel (a) is Flight 5 and panel (b) is Flight 6.

WSW; thus the drainage term varies with height from positive to negative values. The observed wind direction change averages about 15° hr^{-1}, which is not significantly different from the average of $-\partial\delta/\partial t$ over several hundred meters, as shown in Figure 5. For Flight 6, the observed wind direction change is about 3° hr^{-1}, which again is in qualitative agreement with the calculated value of $-\partial\delta/\partial t$ averaged over several hundred meters. In both cases, the friction and drainage terms result in a decrease in $\partial\delta/\partial t$ from a positive value near the surface to a negative value above 200 m, which implies increasing shear with time. These estimates are based on radiosonde and surface measurements at 0600 LT, before sunrise. After sunrise, we expect that the shear would decrease as surface heating increases mixing in the boundary layer. The friction term, which is obtained from the velocity covariances measured during the slant airplane soundings, is positive for both cases. This is consistent with the modeling results of Brost and Wyngaard (1978), for their case with the angle between the surface geostrophic wind and the fall line vector being 180°, which is similar to the values for both Flights 5 and 6. Near the top of the boundary layer, the turbulence, and thus the friction term vanish.

Neumann (1977) has discussed the physical interpretation of the pressure-coriolis term, and pointed out that it can vary even if the large-scale pressure gradient is constant. Basically, it is the difference between the geostrophic parameter and a term involving the scalar product of the pressure gradient and the wind vector.

The horizontal advection term for both cases is negative, and is larger for Flight 5 than Flight 6; for Flight 5, it is comparable to the largest term in the budget for $\partial\delta/\partial t$.

The sloping terrain affects not only the wind direction, but also its magnitude. We now consider the effect of the sloping terrain on the mean wind speed profile; for simplification, we limit our attention to its effect on wind speed near the surface. For $z/h < 0.2$, the change in wind direction was observed to be $< 20°$. Therefore, we need consider only the u-component equation of motion. For steady state, (35) becomes

$$\frac{\partial \overline{uw}}{\partial z} = -\frac{\overline{T}'}{T_0} g\,|\beta| \cos\gamma - \mathrm{f}G \sin\alpha\,, \tag{39}$$

where G is the magnitude of the geostrophic wind at the surface and α is the angle between the geostrophic wind and the surface wind. We assume that

$$\overline{uw} = -K_m \frac{\partial U}{\partial z}\,. \tag{40}$$

Substituting (40) into (39), we have

$$\frac{\partial}{\partial z} K_m \frac{\partial U}{\partial z} = \frac{\overline{T}'}{T_0} g\,|\beta| \cos\gamma + fG \sin\alpha\,. \tag{41}$$

Integrating (41) from z_0 to z, and assuming that the distribution of $\overline{T}'$ with height is

$$\overline{T}' = \langle \overline{T}' \rangle \left(1 - \frac{z}{h_T}\right), \tag{42}$$

where h_T is the height of the inversion – i.e., the height to which significant nocturnal cooling appears to have occurred – (41) becomes

$$K_m \frac{\partial U}{\partial z} = \frac{\langle \overline{T}' \rangle}{T_0} g\,|\beta| \left(z - \frac{z^2}{2h_T} \right) \cos \gamma + fGz \sin \alpha + u_*^2$$

$$= u_*^2 \left(1 + \frac{z}{Y} \right) + \frac{\langle \overline{T}' \rangle}{T_0} g\,|\beta| \left(z - \frac{z^2}{2h_T} \right) \cos \gamma\,, \tag{43}$$

where $Y = ku_* / (fA_1 \sin \alpha)$, and $A_1 = kG/u_*$. The parameter A_1 is a function of stability, and the inversion height h_T is, in general, not the same as the boundary-layer height h, which is the height to which turbulence mixing extends. For early morning flights, the inversion height extends to nearly a kilometer, while the height of turbulent mixing is typically 300 to 500 m. The geostrophic drag law for a flat horizontally homogeneous surface can be written as (Businger and Arya, 1974):

$$(G/u_*) \cos \alpha = k^{-1} (\ln |u_* / fz_0| - A)\,, \tag{44}$$

$$(G/u_*) \sin \alpha = k^{-1} B \operatorname{sgn}(f)\,, \tag{45}$$

where A and B are parameters that depend on stability. Brost and Wyngaard (1978) obtained the following expressions for A and B from their modeling results:

$$A = \ln(h/L) - 2.2(h/L) + 2.9\,, \tag{46}$$

$$B = 3.5(h/L)\,. \tag{47}$$

We assume that the values of A and B change very little from their values at zero slope for a slope of $\beta = 0.003$ (Gutman and Melgarejo, 1981). Therefore, values of A and B were computed from (46) and (47) using the values of h/L for Flights 5 and 6, and substituted into (44) and (45) to estimate G/u_* and α. Using the formulation of Brost and Wyngaard (1978), the steady-state profile of eddy diffusivity for momentum K_m can be expressed as

$$K_m = Dku_* z\,, \tag{48}$$

where $D = (1 - z/h)^{1.5} / [1 + 4.7(z/h)(h/L)]$. Substituting (48) into (43), the normalized wind profile is given by

$$\frac{Dkz}{u_*} \frac{\partial U}{\partial z} = (1 + z/Y) + u_*^{-2} \frac{g}{T_0} \langle \overline{T}' \rangle\, |\beta| \cos \gamma \left(z - \frac{z^2}{2h_T} \right). \tag{49}$$

Introducing the Monin–Obukhov length L into (49), we have

$$\frac{Dkz}{u_*} \frac{\partial U}{\partial z} = 1 + \underbrace{A' \frac{z}{L}}_{\phi_1} + \underbrace{B' \frac{z}{L} \left(1 - 0.5 \frac{z}{h_T} \right)}_{\phi_2}, \tag{50}$$

where $A' = [G/(u_* \mu)] \sin \alpha$, $B' = [\langle \overline{T}' \rangle/(k\theta_*)] \, |\beta| \cos \gamma$, and $\mu = u_*/(|f| \, L)$. Equations (49) and (50) are the normalized wind shear in the lower part of the atmospheric boundary layer over sloping terrain, and ϕ_1 and ϕ_2 are universal functions for the normalized shear which represent, respectively, the effects of stratification over flat terrain and the contribution due to sloping terrain. Figure 16 shows the distributions of ϕ_1 and ϕ_2 with normalized height z/h_T and Figure 17 shows the variation of ϕ_1 and ϕ_2 with z/L for Flights 5 and 6.

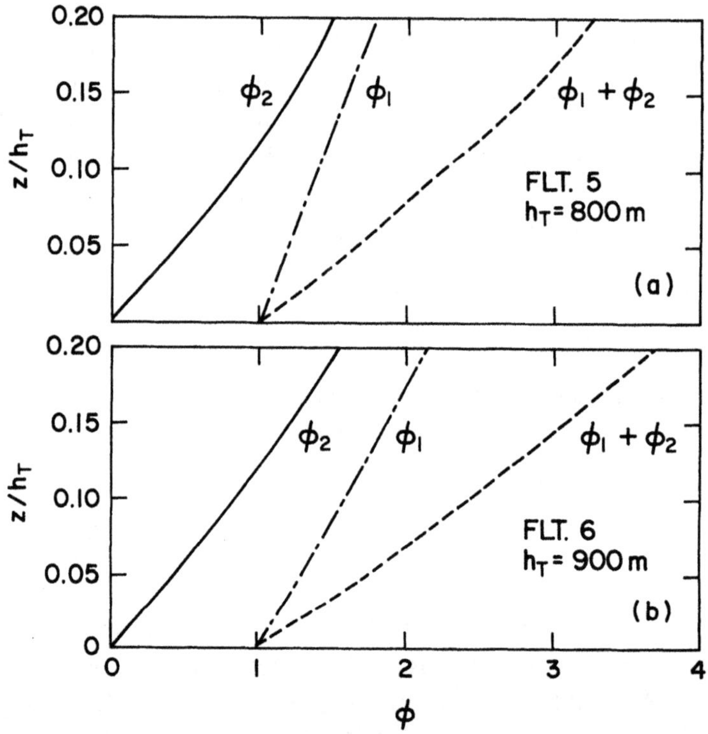

Fig. 16. Normalized wind shear. The function ϕ_1 is the contribution due to stratification and ϕ_2 is the contribution due to terrain slope.

We see that for Flights 5 and 6 the effect of terrain slope is to increase the wind shear by about 25 to 30% compared to the flat-surface value at a height of $z/h_T = 0.05$, which is about the lower limit for estimating wind shear from the airplane measurements. For Flight 5, the measured value of $\partial U/\partial z$ at $z/h_T = 0.05$ was 0.04 s^{-1} and the value estimated from (50) is 0.03 s^{-1}; for Flight 6, the corresponding values are 0.03 s^{-1} and 0.04 s^{-1}, respectively.

3.3. Turbulence Energy Budget

The turbulence budget over sloped terrain can be written as

$$\frac{\partial \overline{e}}{\partial t} = -\overline{uw} \frac{\partial U}{\partial z} - \overline{vw} \frac{\partial V}{\partial z} - \frac{\partial(\overline{we} + \overline{pw}/\rho_0)}{\partial z} + \\ + \frac{g}{\Theta_0} (\overline{w\theta} - |\beta| \, \overline{\theta u} \cos \gamma + |\beta| \, \overline{\theta v}) \cos \gamma - \varepsilon \, , \tag{51}$$

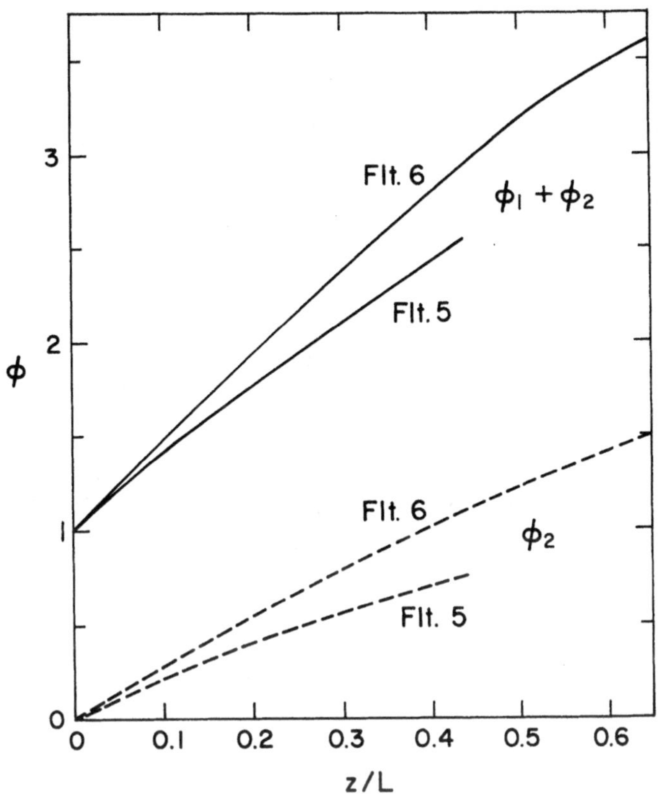

Fig. 17. Normalized wind shear as a function of z/L. The function ϕ_1 is the contribution due to stratification and ϕ_2 that due to terrain slope.

where ε is the viscous dissipation of turbulence energy, p is pressure, and the other notation is standard. The dissipation is obtained from the inertial subrange of the velocity spectra. This was done by first highpass filtering the velocity components with a 3 Hz cutoff filter ($\sim$ 23 m wavelength) to ensure that only the inertial subrange is included in the estimate, then processing the filtered data as was done for other second-moment variables; i.e., squaring the filtered output, and finally lowpass filtering the result (Figure 2).

We calculated all the terms in the turbulence energy equation, except for the pressure transport term, for four of the SESAME cases using data from the aircraft slant flight paths; Figure 18 is typical. In all cases, the time variations and the terrain slope terms are negligible, and the dominant terms are shear production and viscous dissipation, with a significant residual, which, when integrated over the entire boundary layer, is negligible. The terms decrease approximately exponentially with height through the boundary layer, but in two cases (Flights 4 and 6) there is a tendency for the shear production to increase near the level of enhanced temperature inversion, about twice the height of the boundary layer defined by surface-based turbulence. We suspect that this is mainly wave-induced activity.

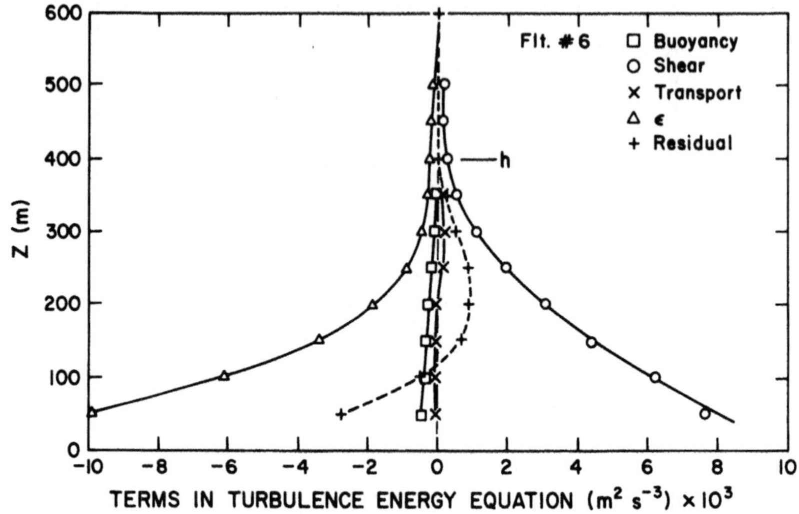

Fig. 18. Turbulence kinetic energy budget for Flight 6.

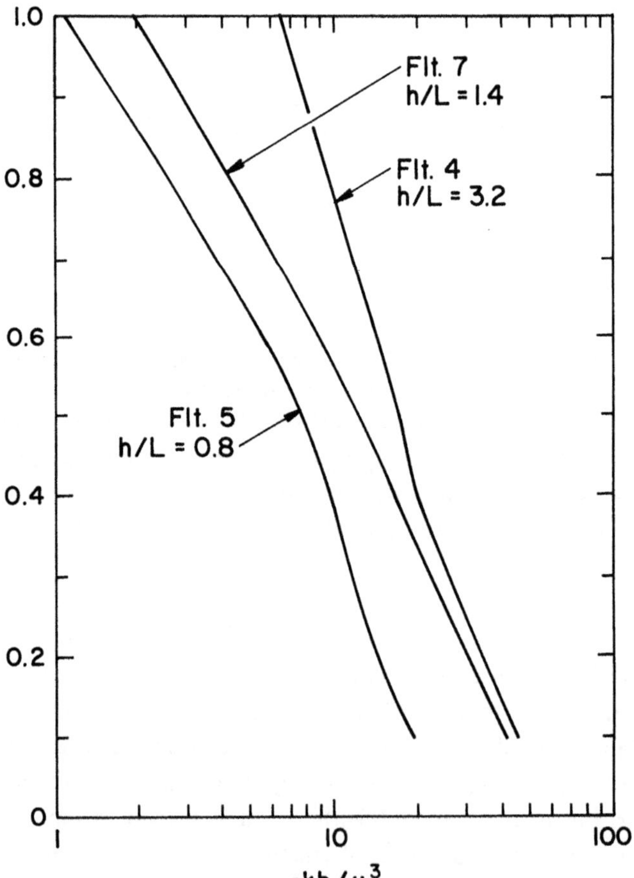

Fig. 19. Normalized dissipation for Flights 4, 5, and 7.

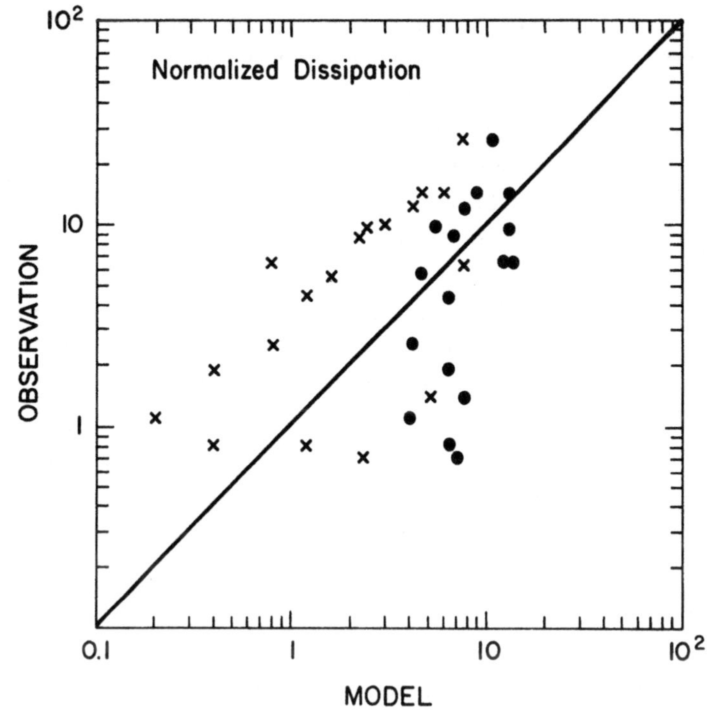

Fig. 20. A comparison of observed values of normalized dissipation with that obtained from the model of Brost and Wyngaard (1978). The ×'s are values observed here and the solid circles are values reported by Caughey *et al.* (1979).

We have plotted the normalized turbulence energy dissipation rate $\varepsilon kh/u_*^3$ versus z/h in Figure 19 for the slant flight paths of Flights 4, 5, and 7. We then compare these results in Figure 20 with the parameterization developed by Brost and Wyngaard (1978),

$$\frac{\varepsilon kh}{u_*^3} = \left(\frac{h}{z}\right)\left[1 + 3.7\left(\frac{z}{h}\right)\left(\frac{h}{L}\right)\right]\left(1 - 0.85\,\frac{z}{h}\right)^{1.5}. \tag{52}$$

The observation points in Figure 20 are the values shown in Figure 19 at intervals of $z/h = 0.2$, as well as the results of Caughey *et al.* (1979). The agreement is generally good, although our observations tend to be somewhat greater than the predicted values; the agreement is somewhat better for our observations than for those reported by Caughey *et al.* (1979).

4. Concluding Remarks

We have shown that, for the most part, the mean and turbulent structure of the NBL over the Great Plains obtained from aircraft measurements during SESAME agrees with previous tower-based and tethered balloon observations and with NBL similarity theory. [In contrast, the vertical structure for the very stable case on the morning of

5 May, 1979 (Mahrt, 1985) was much different; the principal source of turbulence was shear generation at the top of the surface inversion layer.] The major difference is the relatively small value of K_h/K_m obtained here, which results in a smaller value of Ri_f as compared to previous studies. We speculate that this may be a result of the relatively rougher terrain encountered here. In Part II, we show evidence for standing waves that may exist over the experimental area, which may contribute to the momentum flux measured by the aircraft, but not to tower or tethered-balloon observations.

The results show that aircraft are a useful tool for probing the NBL. Concurrent aircraft and tower or tethered balloon measurements of mean and turbulence structure would be even more useful. The aircraft can observe the spatial structure, while the stationary platforms can obtain longer time series and observations closer to the surface. Airplane measurements could also be considerably enhanced by direct measurements of the Richardson number – i.e., direct measurements of the wind shear and lapse rate – and thus directly relate the mean structure to the turbulence intensity. As pointed out by Keeler *et al.* (1987) and Kristensen and Lenschow (1987), this now seems technically feasible.

Acknowledgements

We thank G. D. Emmitt for providing us with the acoustic sounder records during the experimental period, and the personnel at the NCAR Research Aviation Facility for their assistance during and after the experiment. We also thank C. Nappo, J. Wyngaard, L. Mahrt, Z. Sorbjan, and M. Raupach for their helpful comments.

References

André, J. C. and Mahrt, L.: 1982, 'The Nocturnal Surface Inversion and Influence of Clear-Air Radiative Cooling', *J. Atmos. Sci.* **39**, 864–878.

André, J. C., De Moor, G., Lacarrère, P., Therry, G., and du Vachat, R.: 1978, 'Modeling the 24-hour Evolution of the Mean and Turbulent Structures of the Planetary Boundary Layer', *J. Atmos. Sci.* **35**, 1861–1883.

Bonner, W. D., Steven, D. E., and Robert, G.: 1968, 'Kinematics of the Low-Level Jet', *J. Appl. Meteorol.* **7**, 339–347.

Brost, R. A. and Wyngaard, J. C.: 1978, 'A Model Study of the Stably Stratified Planetary Boundary Layer', *J. Atmos. Sci.* **35**, 1427–1440.

Businger, J. A. and Arya, S. P. S.: 1974, 'Height of the Mixed Layer in the Stably Stratified Planetary Boundary Layer', *Adv. Geophys.* **18a**, Academic Press, New York, 73–92.

Caughey, S. J., Wyngaard, J. C., and Kaimal, J. C.: 1979, 'Turbulence in the Evolving Stable Layer', *J. Atmos. Sci.* **36**, 1041–1052.

Garratt, J. R.: 1982, 'Observations in the Nocturnal Boundary Layer', *Boundary-Layer Meteorol.* **22**, 21–48.

Garratt, J. R. and Brost, R. A.: 1981, 'Radiative Cooling Effects Within and Above the Nocturnal Boundary Layer', *J. Atmos. Sci.* **38**, 2730–2746.

Gossard, E. E. and Hooke, W. H.: 1975, *Waves in the Atmosphere*, Elsevier Scientific Publ. Co., New York, p. 456.

Gutman, L. N. and Melgarejo, J. W.: 1981, 'On the Laws of Geostrophic Drag and Heat Transfer Over a Slightly Inclined Terrain', *J. Atmos. Sci.* **38**, 1714–1724.

Keeler, R. J., Serafin, R. J., Schwiesow, R. L., Lenschow, D. H., Vaughan, J. M., and Woodfield, A.: 1987, 'An Airborne Laser Air Motion Sensing System. Part I: Concept and Preliminary Experiment', *J. Atmos. and Oceanic Tech.* **4**, 113–127.

Kondo, J., Kanechika, O., and Yasuda, N.: 1978, 'Heat and Momentum Transfers Under Strong Stability in the Atmospheric Surface Layer', *J. Atmos. Sci.* **35**, 1012–1021.

Kristensen, L. and Lenschow, D. H.: 1987, 'An Airborne Laser Air Motion Sensing System. Part II: Design Criteria and Measurement Possibilities', *J. Atmos. and Oceanic Tech.* **4**, 128–138.

Lacser, A. and Arya, S. P. S.: 1986, 'A Numerical Model Study of the Structure and Similarity Scaling of the Nocturnal Boundary Layer (NBL)', *Boundary-Layer Meteorol.* **35**, 369–385.

Lenschow, D. H.: 1986, 'Aircraft Measurements in the Boundary Layer', in D. H. Lenschow (ed.), *Probing the Atmospheric Boundary Layer*, American Meteorological Society, Boston, MA, pp. 39–55.

Lenschow, D. H. and Stephens, P. L.: 1980, 'The Role of Thermals in the Convective Boundary Layer', *Boundary-Layer Meteorol.* **19**, 509–532.

Lenschow, D. H. and Stankov, B. B.: 1986, 'Length Scales in the Convective Atmospheric Boundary Layer', *J. Atmos. Sci.* **43**, 1198–1209.

Mahrt, L., Heald, R. C., Lenschow, D. H., and Stankov, B. B.: 1979, 'An Observational Study of the Structure of the Nocturnal Boundary Layer', *Boundary-Layer Meteorol.* **17**, 247–264.

Mahrt, L.: 1985, 'Vertical Structure and Turbulence in the Very Stable Boundary Layer', *J. Atmos. Sci.* **42**, 2333–2349.

Neumann, J.: 1977, 'On the Rotation Rate of the Direction of the Sea and Land Breezes', *J. Atmos. Sci.* **34**, 1913–1917.

Nieuwstadt, F. T. M.: 1980, 'A Rate Equation for the Inversion Height in a Nocturnal Boundary Layer', *J. Appl. Meteorol.* **19**, 1445–1447.

Nieuwstadt, F. T. M.: 1984, 'The Turbulent Structure of the Stable, Nocturnal Boundary Layer', *J. Atmos. Sci.* **41**, 2202–2216.

Nieuwstadt, F. T. M.: 1985, 'A Model for the Stationary, Stable Boundary Layer', in J. R. C. Hunt (ed.), *Turbulence and Diffusion in Stable Environments*, Oxford University Press, pp. 149–179.

Paegle, J. and Resch, G. E.: 1973, 'Three-Dimensional Characteristics of Diurnally Varying Boundary-Layer Flows', *Monthly Weather Rev.* **101**, 746–756.

Pasquill, F. and Smith, F. B.: 1983, *Atmospheric Diffusion*, 3rd ed., Ellis Horwood Publishers, England.

Sorbjan, Z.: 1987, 'Comments on 'Scaling the Atmospheric Boundary Layer'', *Boundary-Layer Meteorol.* **38**, 411–413.

Ueda, H., Mitsumoto, S., and Komori, S.: 1981, 'Buoyancy Effects on the Turbulent Transport Processes in the Lower Atmosphere', *Quart. J. R. Meteorol. Soc.* **107**, 561–578.

Wyngaard, J. C., Pennell, W. T., Lenschow, D. H., and LeMone, M. A.: 1978, 'The Temperature-Humidity Covariance Budget in the Convective Boundary Layer', *J. Atmos. Sci.* **35**, 47–58.

Yamada, T.: 1979, 'Prediction of the Nocturnal Surface Inversion Height', *J. Appl. Meteorol.* **18**, 526–531.

Zilitinkevich, S. S.: 1972, 'On the Determination of the Height of the Ekman Boundary Layer', *Boundary-Layer Meteorol.* **3**, 141–145.

THE STABLY STRATIFIED BOUNDARY LAYER OVER THE GREAT PLAINS

II. *Horizontal Variations and Spectra*

DONALD H. LENSCHOW, SHI F. ZHANG[1], and B. BOBA STANKOV[2]

National Center for Atmospheric Research, P.O. Box 3000, Boulder, CO 80307, U.S.A.*

(Received in final form 29 August, 1987)

Abstract. Airplane measurements of the stably stratified boundary layer obtained during the Severe Environmental Storms and Mesoscale Experiment (SESAME) over rolling terrain in south-central Oklahoma indicate that considerable horizontal variability exists in the flow on scales of several kilometers. Much of this wave-like structure appears to be tied to the terrain. The criteria for existence of stationary gravity waves indicate that these waves can exist under the observed conditions. The spectrum of terrain variations also supports the existence of these waves. Observed spectra of the vertical velocity have two peaks: one at wavelengths of several kilometers, which is due to waves and the other at wavelengths of about 100 m, which is due to turbulence. The variance at several kilometers wavelength increases somewhat with height at least up to about 800 m, but the variance contributed by turbulence decreases rapidly with height.

1. Introduction

The mean and turbulence structure of the Great Plains nocturnal boundary layer (NBL) were discussed in Part I. One characteristic of the NBL, as compared to the convective boundary layer, is its greater sensitivity to variations in terrain. The effects of a mean slope were discussed in Part I; here we discuss the effects of smaller scale variations in terrain, which seem to be responsible for the wave structures observed in the SESAME NBL.

The SESAME flights were conducted in periods of sufficient sunlight to permit low-level flight legs. As a result, they span periods of transition from the daytime to the nighttime boundary layer and from the nighttime to daytime boundary layer. Here we discuss two early morning flights in a typical clear late spring weather situation with moderately strong southerly winds aloft, which veered to the southwest near the nocturnal inversion (Figure 5, Part I). Flight 5 occurred on 6 May, 1979 and Flight 6 on 7 May, 1979. The flight legs were oriented in a north-south direction, and so were approximately parallel to the wind at the lower flight levels. Further details on the experiment and of the mean and turbulence structure of the boundary layer during these flights are presented in Part I. In contrast, Mahrt (1985) has analyzed data from Flight 3, an early morning flight on 5 May with light winds – thus a much more stably stratified boundary layer than discussed here.

[1] Present affiliation: University of Washington, Seattle, WA 98195, U.S.A.
[2] Present affiliation: Meteorological Wave Propagation Laboratory, NOAA, Boulder, CA 80303, U.S.A.
* The National Center for Atmospheric Research is sponsored by the National Science Foundation.

Boundary-Layer Meteorology **42** (1988) 123–135.

In addition to investigating the wave structures, we also discuss the behavior of the velocity spectra and cospectra of vertical velocity and temperature during the nocturnal to daytime transition for one of the early morning cases.

2. Spatial Variability

One of the advantages of aircraft measurements over measurements from a fixed location is the ability to resolve spatial variations. This is of special importance in the nocturnal boundary layer since the terrain has such a strong influence on the boundary-layer structure. One of the most obvious characteristics of the NBL is the presence of both wave motions and turbulence. We show here that, on the one hand, wave amplitudes and wavelengths appear to be relatively constant with height, with some indication that the amplitudes increase with height, particularly near the level of the wind maximum. The turbulence, on the other hand, decreases approximately exponentially and becomes increasingly intermittent with height. Figure 1 shows an example of a

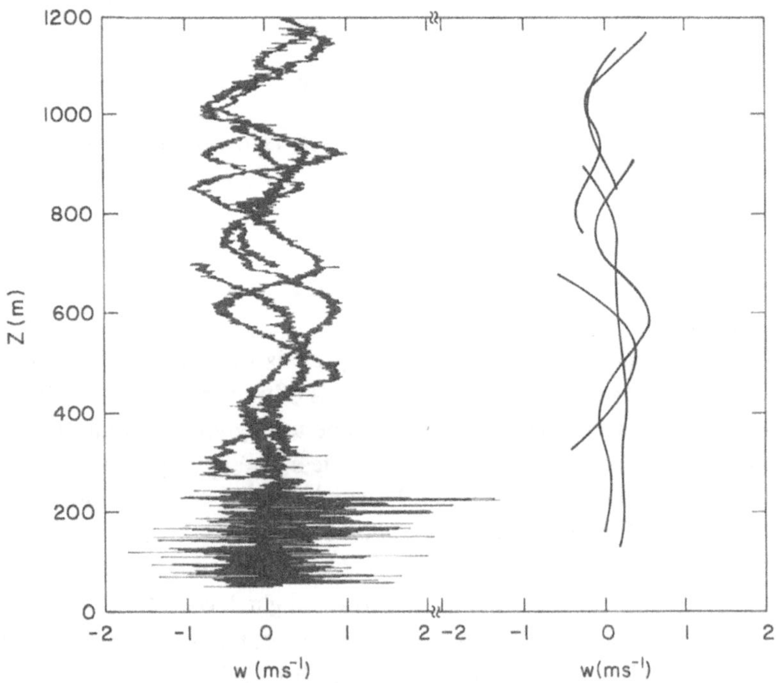

Fig. 1. Vertical soundings of vertical velocity obtained from Flight 6. Left panel shows the unfiltered signal; right panel the low-pass filtered signal.

vertical sounding of vertical velocity without smoothing and with smoothing (with a 0.01 Hz low-pass filter as discussed in Part I), which illustrates the decrease of turbulence with height and the dominance of wave activity above the turbulent layer.

Within the turbulent layer, the smoothed vertical velocity shows little evidence of wave activity. This figure also shows the effect of the averaging process used here. We see that the amplitude of the filtered wave is reduced to about half the unfiltered wave.

Other variables also show evidence of wave activity. Figure 2 shows an example from

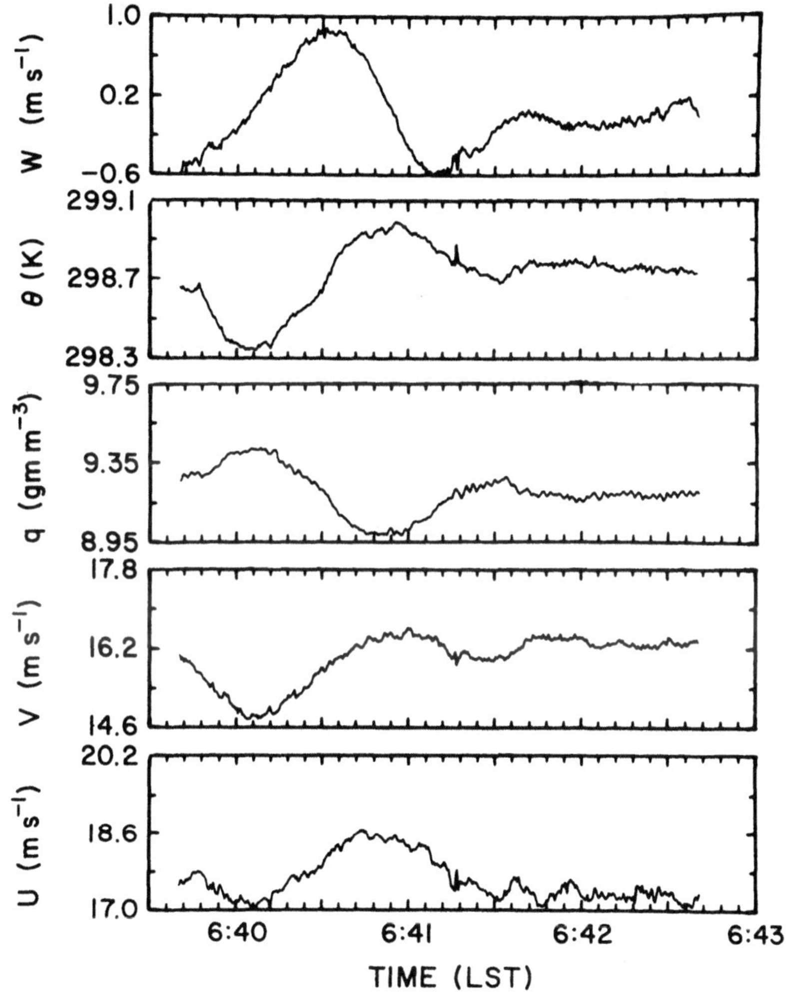

Fig. 2. Time series for a constant level leg 600 m above the ground heading south on Flight 6.

Flight 6 of a constant-level leg at a height of about 600 m, well above the turbulent boundary layer. All three velocity components (u is east, v is north, and w is up), potential temperature θ, and humidity q have similar shapes, although with different phase relationships. The magnitude of the phase angles between w and θ, q, u, and v are $\sim 90°$. Wavelengths of variables were estimated by multiplying the airplane true airspeed by the measured period. We found that the wavelength of the largest amplitude contribution to the variances was shorter on the downwind legs than on the upwind legs, which indicates that at least some of the observed wave activity is tied to the terrain.

Spectra of w at three levels for three time periods during Flight 6 are shown in Figure 3a, and of u and v in Figure 3b. The period of maximum wave activity is at 6 : 27 LT at 90 m and at 6 : 32 LT at 200 m; the w spectral peaks are at ~ 7.1 km at 90 m, which was an upwind leg, and ~ 4.5 km at 200 m, which was a downwind leg. The calculated mean wind speed which would give us this wavelength difference is ~ 15 m s^{-1}, which agrees well with the observed mean wind speed. The general decrease

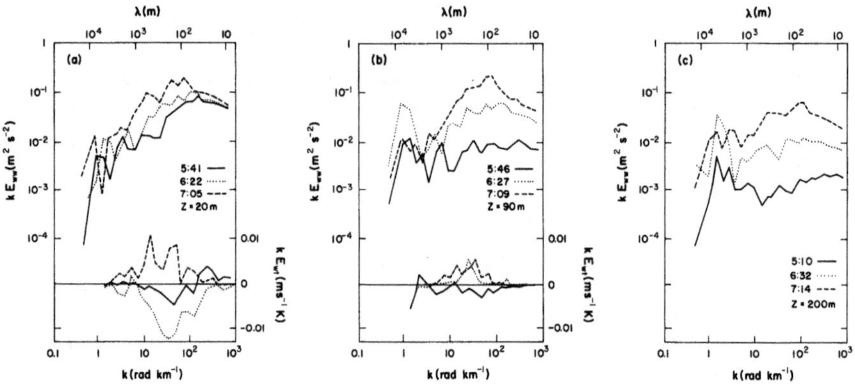

Fig. 3a. Spectra of vertical velocity and cospectra of vertical velocity and temperature for Flight 6 at 20, 90 and 200 m.

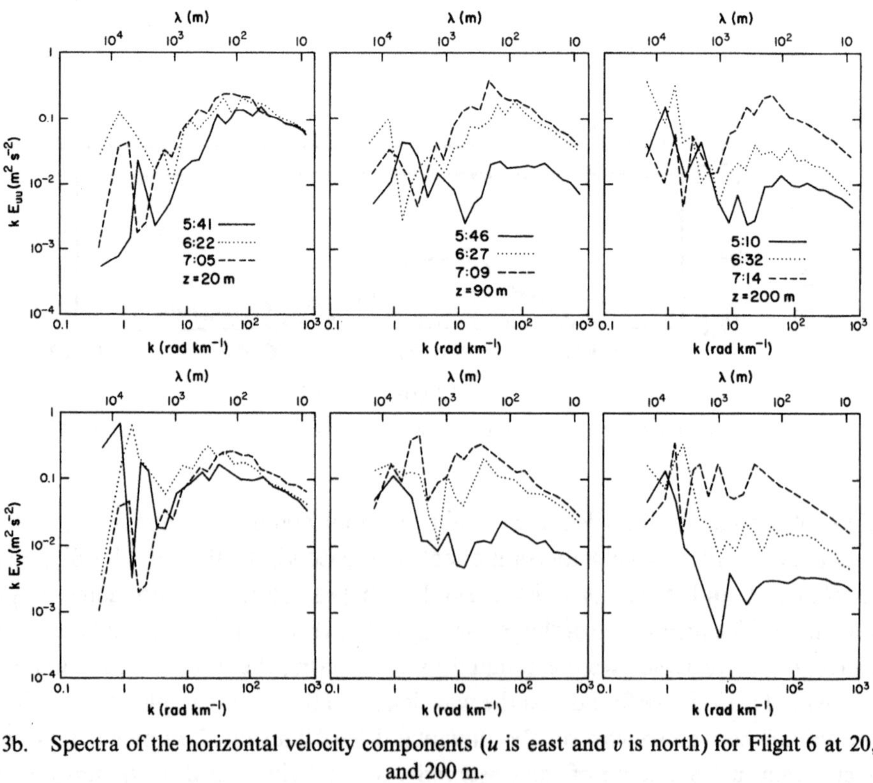

Fig. 3b. Spectra of the horizontal velocity components (u is east and v is north) for Flight 6 at 20, 90, and 200 m.

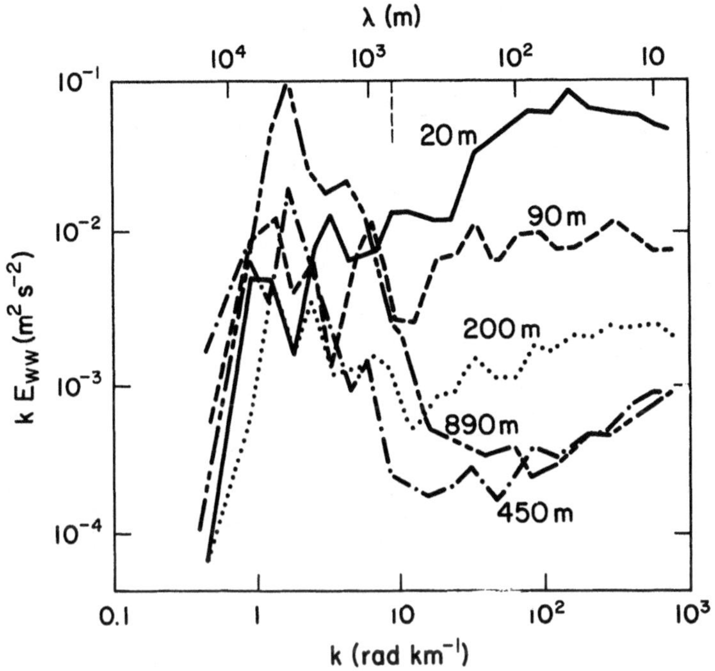

Fig. 4. Spectra of vertical velocity at 5 levels for Flight 6 before 05 : 50 LT. The short dashed vertical line down from the top is the wavelength of the high-pass filter used to filter out the wave contributions to the spectra.

in turbulence amplitude with height and the increase of wave activity is shown more directly in Figure 4.

Even more evidence for the existence of stationary waves on both Flights 5 and 6 was obtained from cross-sections of w, θ, and the wind speed, which were constructed from a composite of the constant-level and slant-path flight legs. The cross-sections from Flights 5 and 6 are shown in Figure 5. During the two to three hours that it took to fly these legs, the fields shown in Figure 5 maintained the observed pattern, which indicates that the wavelike structures were nearly stationary.

We now examine the likely structure of the waves and their relationship to the underlying terrain. We start with the relation for two-dimensional flow in a stably stratified incompressible fluid (e.g., Gossard and Hooke, 1975)

$$W_{zz} + \left[\frac{N^2 k^2}{(\omega - Uk)^2} + \frac{kU_{zz}}{(\omega - Uk)} - k^2\right] W = 0\,, \tag{1}$$

where W is the amplitude of the vertical wave velocity, $N = [(g/\theta)(\partial\theta/\partial z)]^{1/2}$ is the Brunt–Väisälä frequency, k is wave number, ω is angular frequency, and differentiation with respect to z is denoted by a subscript z.

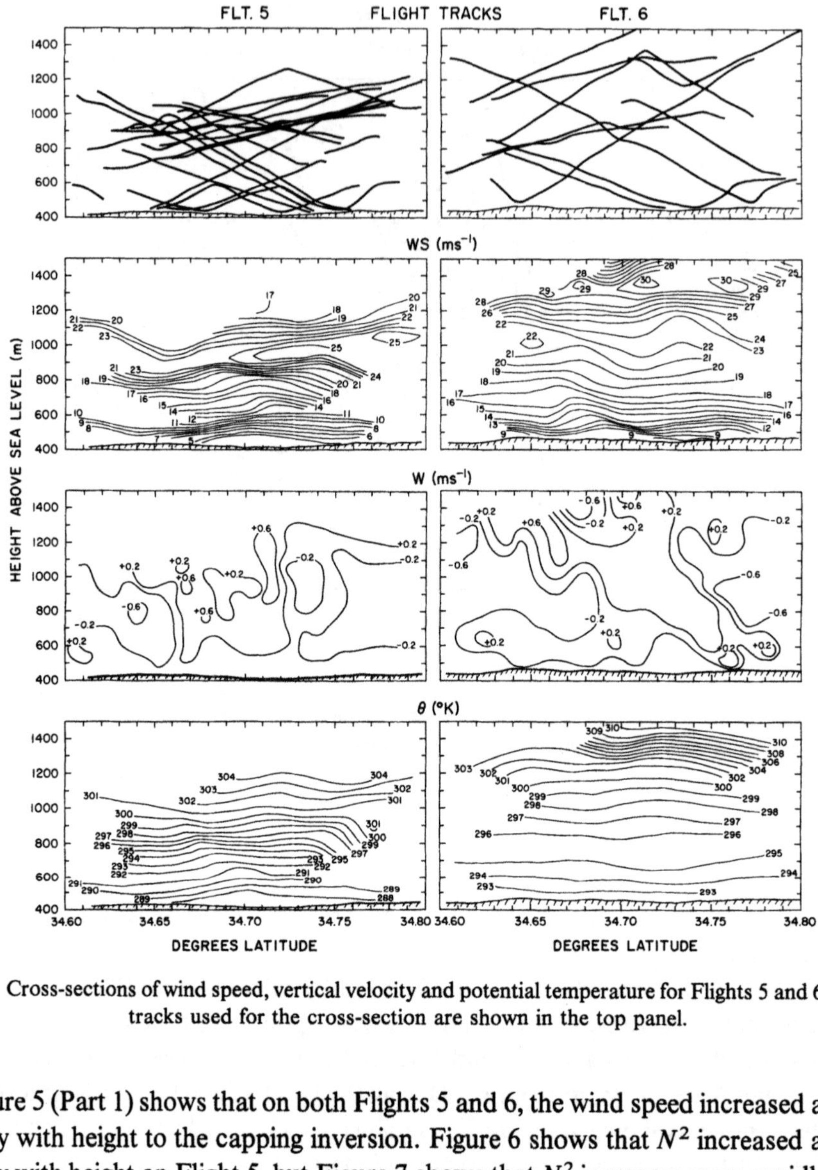

Fig. 5. Cross-sections of wind speed, vertical velocity and potential temperature for Flights 5 and 6. Flight tracks used for the cross-section are shown in the top panel.

Figure 5 (Part 1) shows that on both Flights 5 and 6, the wind speed increased almost linearly with height to the capping inversion. Figure 6 shows that N^2 increased almost linearly with height on Flight 5, but Figure 7 shows that N^2 increases more rapidly than this on Flight 6. This means that the potential temperature profile has greater curvature on Flight 6 than on Flight 5. Table I shows the wind speed and N^2 profiles assumed for this analysis. The behavior of N^2 is quite different from that used in the analysis of Rotunno (1977), who studied the behavior of internal gravity waves in a time-varying stratification. He found that for the Wangara Experiment (Clarke *et al.*, 1971) N^2 decreased approximately exponentially with height; i.e., the curvature in the temperature profile was observed to decrease with height, while here it increased with height up to the inversion. Similarly, Hunt *et al.* (1985) examined three cases at the Boulder Atmospheric Observatory (BAO) which also showed an approximate exponential decrease with height. As pointed out by André and Mahrt (1982), negative curvature

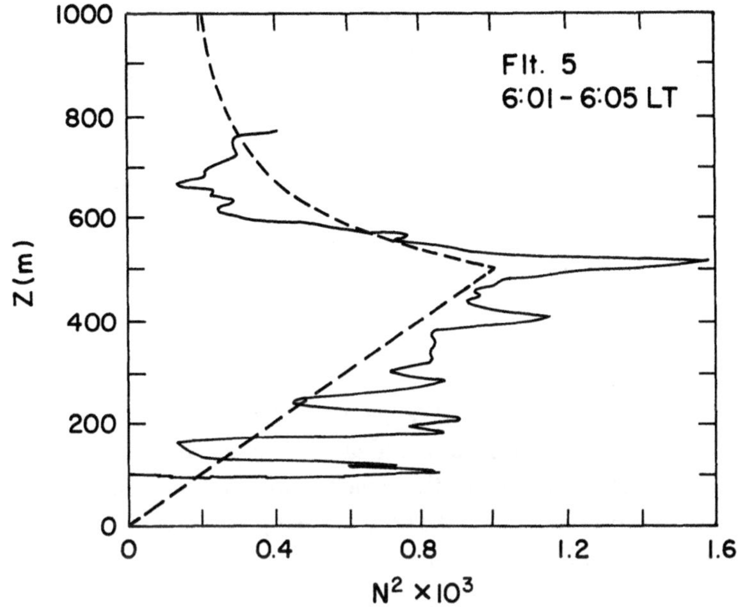

Fig. 6. Profile of measured (solid line) and modeled (dashed line) N^2 for Flight 5. The measured profile has been smoothed with a 0.02 Hz low-pass filter.

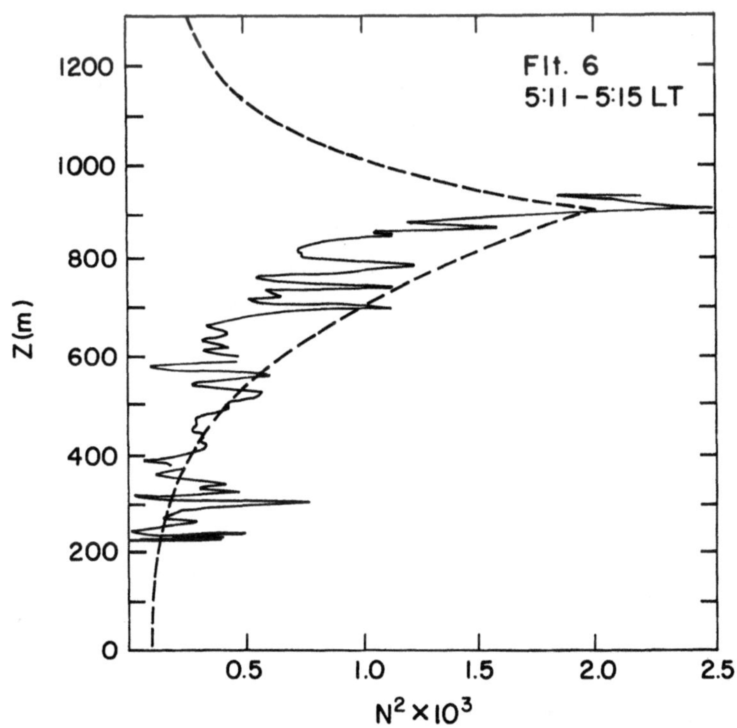

Fig. 7. Same as Figure 6 for Flight 6.

TABLE I

Profiles of N^2 and U assumed for calculating wave characteristics for Flights 5 and 6

Flt. No.	h (m)	Height interval (m)	N^2 (s^{-2})	U (m s^{-1})
5	500	$0 < z < 500$	$2 \times 10^{-6} z$	$7 + 0.032z$
		$z > 500$	$0.0008 \exp^{(z-500)/125} + 0.0002$	23
6	900	$0 < z < 900$	$0.0019(z/900)^3 + 0.0001$	$10 + z/45$
		$z > 900$	$0.0018 \exp^{-(z-900)/125} + 0.0002$	30

of the temperature profile, as in the Wangara and BAO NBLs, is an indication that sensible heat transport is dominated by radiation divergence; while positive curvature is an indication that turbulence is relatively more important for sensible heat transport. This agrees with the observation that the potential temperature profile for Flight 6, which had higher wind speeds, has more curvature than for Flight 5.

Since the observations indicate the presence of standing waves, we set $\omega = 0$ in (1); thus,

$$W_{zz} + l^2 W = 0 , \tag{2}$$

where $l^2 \equiv (N^2/U^2 - U_{zz}/U - k^2)$. Inserting the values of N^2 and wind speed from Table I into the expression for l, we find that U_{zz}/U is small compared to the other two terms throughout most of the boundary layer. Hunt *et al.* (1985) found that this term was larger than the buoyancy term in the BAO boundary layer below 100 m. In the cases presented here, the measurement resolution below 100 m is not sufficient to estimate the velocity curvature term, but extrapolating the observed wind profiles down to the surface indicates that this term is also important near the surface.

If we assume, for the moment, that we can approximate the NBL by a two-layer model such that l^2 is constant up to a height h and zero above h (here h represents the height of the inversion, not the height of the turbulent boundary layer; in Part I, this height is denoted by h_T), and that $W(0) = W(h) = 0$, we have for the solution to (2) $W = W(0) \sin lz$, with $hl = n\pi$, where n is an integer. Therefore,

$$k^2 = \frac{N^2}{U^2} - \frac{(n\pi)^2}{h^2} . \tag{3}$$

In order for lee waves to exist, $k^2 > 0$. Putting in approximate observed mean values for N^2 and U, and $n = 1$, this seems not to be the case for either of the two cases considered here. In effect, the mean wind speed is strong enough that the phase velocity of the waves is less than the mean wind speed at all levels, and consequently lee waves cannot exist.

We, therefore, consider the local response of the atmosphere to the underlying terrain by numerically solving (2) using the values of N^2 and wind speed given in Table I, and the lower boundary condition

$$W(0) = -iU_0 k_0 H(k), \tag{4}$$

where $H(k)$ is the Fourier transformed terrain height in the direction of the wind vector,

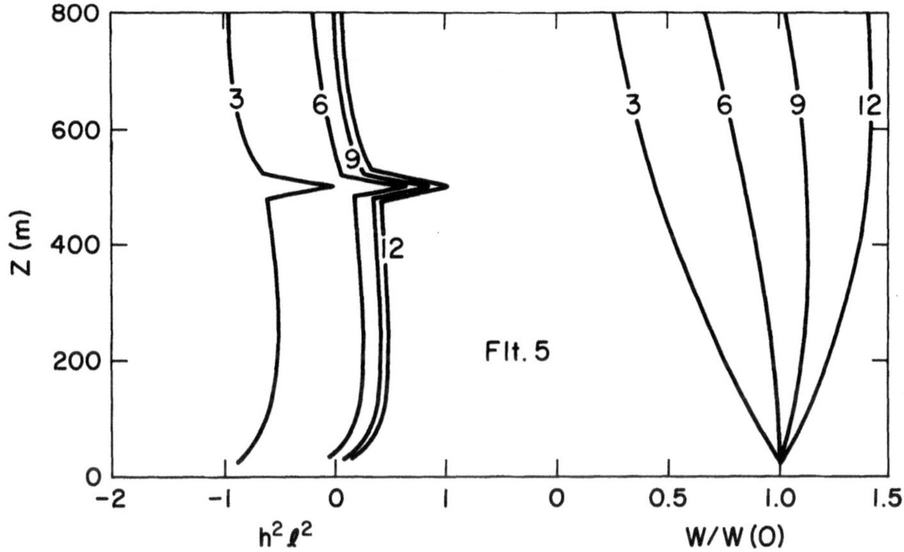

Fig. 8. Calculated profiles of $h^2 l^2$ and wave amplitude for Flight 5 for wavelengths of 3, 6, 9, and 12 km.

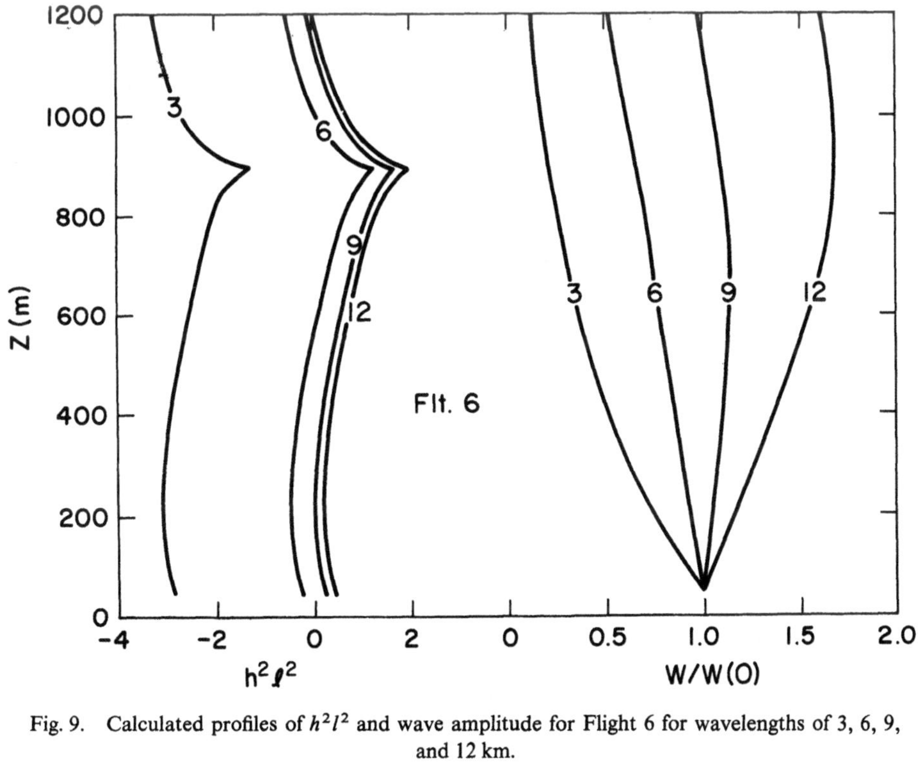

Fig. 9. Calculated profiles of $h^2 l^2$ and wave amplitude for Flight 6 for wavelengths of 3, 6, 9, and 12 km.

and the 0 subscript denotes surface values. The numerical calculations were carried out iteratively to obtain $W = 0$ at $z = 20$ km, using a height increment of $\Delta z = 25$ m. Values of $z > 20$ km and $\Delta z < 25$ m did not significantly change the computed profiles of W below 2 km. Numerical calculations were also compared with analytic solutions of two-layer models of the atmosphere, with good agreement. Figures 8 and 9 show solutions of (2) for wavelengths $\lambda = 2\pi/k = 3, 6, 9$, and 12 km. We see that indeed for wavelengths greater than about 8 km, the wave amplitude increases with height up to the inversion, and for wavelengths smaller than this the amplitude is damped with height.

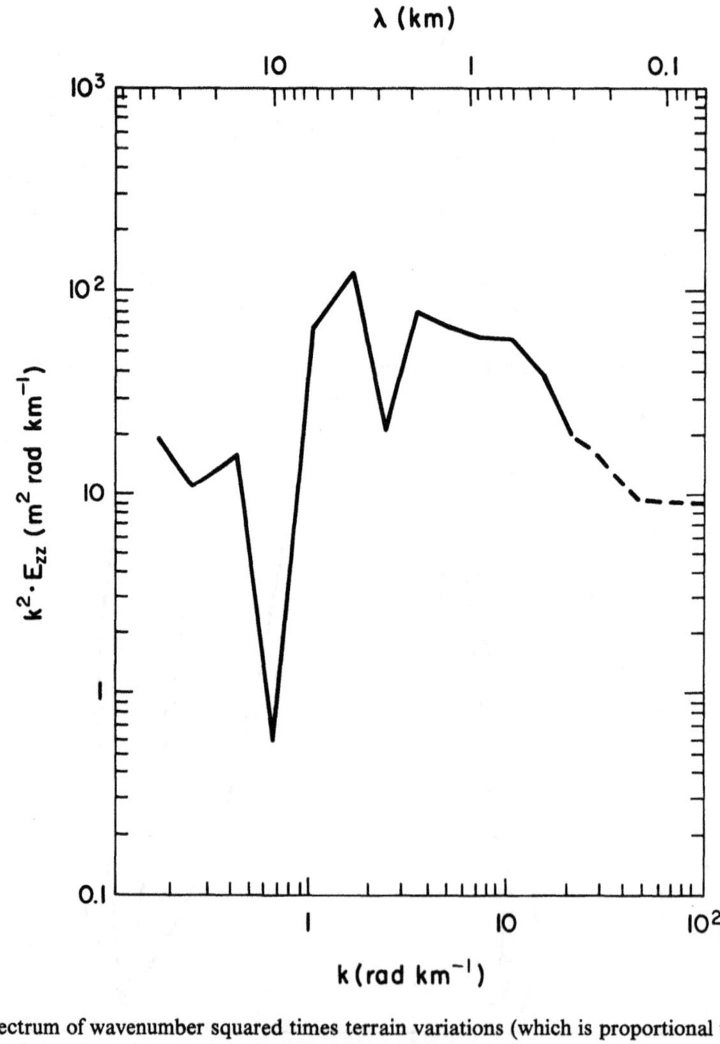

Fig. 10. Spectrum of wavenumber squared times terrain variations (which is proportional to the vertical velocity component of air flow over the terrain) in the measurement area obtained from airplane measurements of pressure altitude and geometric altitude. The dashed portion of the curve is geometric altitude only; at high wavenumbers the pressure altitude signal is noisy, but the aircraft vertical velocity is small, so that the contribution of pressure altitude to the output is negligible.

Figure 10 is a plot of the spectrum of terrain height multiplied by wavenumber squared, which is obtained from airplane measurements of the geometric altitude (measured with a radar altimeter) minus the pressure altitude (measured with a pressure altimeter). This curve is proportional to the spectrum of the vertical air velocity at the surface (4), assuming a constant wind speed. The spectral peak is in the range of 4 to 7 km; above this, the variance drops off drastically. Integrating the spectrum over the range of $1 < \lambda < 10$ km, and assuming a surface wind speed U_0 of 10 m s^{-1}, we obtain a value of $W(0) \sim 0.25$ m s^{-1}. This is of similar magnitude to the values obtained by integrating the vertical velocity spectra over the same wavelength region for those cases with a pronounced peak in the region of 1 to 10 km wavelengths. Thus, the observed stationary wave structures agree well with those predicted based on flow of a stably stratified inversion-capped fluid over rolling terrain.

The observed spectra, as shown for example in Figures 3 and 4, also indicate that the wave activity increases with height. At 20 m, there is little or no indication of wave activity, while at 90 and 200 m, the spectral density in the range of 1 to 10 km is larger than near the surface. The model does not predict significant amplification of the waves with height for $\lambda < 9$ km, which may be a consequence of an oversimplified model.

3. Time Variability

One of the limitations in the vertical velocity spectra shown in Figures 3 and 4 is that at 200 m and higher, the variance in the high wavenumber portion of the spectra at the earliest time before solar heating occurs is so small that the results are contaminated by noise. Therefore, it is difficult to estimate the location of the spectral peak at these heights. Generally, however, we find the spectral peaks at somewhat longer wavelengths than predicted by Caughey *et al.* (1979). (We also point out that noise in the spectra tends to decrease the wavelength of the spectral peak.) At 20 m, for example, the normalized vertical profiles that Caughey *et al.* (1979) found from their data (over flat terrain in northern Minnesota) predict a peak in the w spectra for Flight 6 at about 20 m wavelength; we find here that the peak is at about 50 m. Similarly, they predict a spectral peak in the cospectrum of w and θ at about 40 m wavelength, while we find the peak at about 100 m. We suggest that this may again be due to the rougher terrain present in our case, although we cannot exclude the possibility that spatially averaged spectra may differ somewhat from temporally averaged spectra in the short wavelength end of the spectra, since they are likely to differ considerably at longer wavelengths.

There is, however, an obvious increase in scale and amplitude with time, as heating begins to occur after sunrise. The effect also increases with height; the amplitude increase is about a factor of 5 for wavelengths > 100 m and is almost unnoticeable for wavelengths < 20 m at 20 m height; at 200 m the spectra increases by about a factor of 30 in the one and a half hours of measurement. The spectral peak increases in wavelength from about 50 m to about 100 m over this time interval, and remains almost constant with height. We note that the increase is not large enough to result in greater turbulence dissipation at the upper levels compared to the surface. The inertial subrange spectral region is still larger at 20 m than at the upper levels.

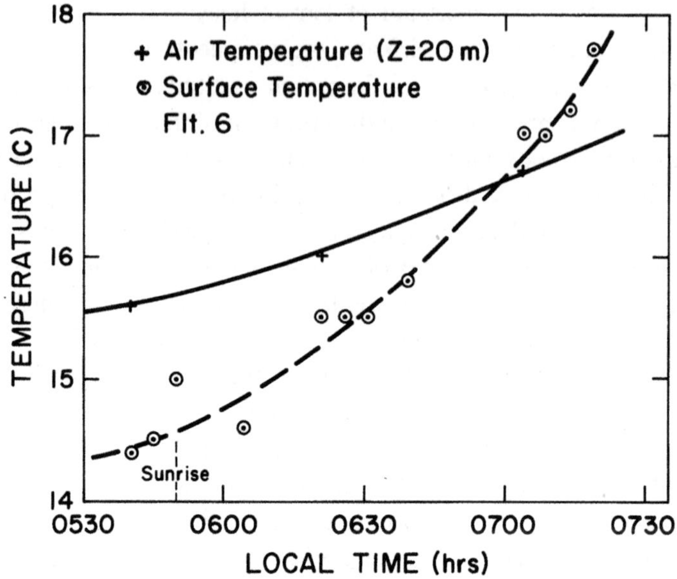

Fig. 11. Time series of the radiation surface temperature and the air temperature measured by the aircraft at 20 m height for Flight 6.

Figure 11 shows that the surface temperature, obtained from a downward-looking Barnes PRT-5 Surface Temperature Radiometer, begins to rise within a half-hour after sunrise, and becomes warmer than the temperature at 20 m height about an hour after sunrise. The sensible heat flux (Figure 3) initially decreases after sunrise, which is likely a result of the increased mixing that follows the reduced stabilization near the surface. This leads to a local cooling transient in the layer $20 > z > 90$ m as the warmer air is mixed downward by the enhanced turbulence. Within an hour after sunrise, however, the heat flux is already positive at 90 m. By 0700 LT, contributions to the spectra by waves are no longer evident.

4. Concluding Remarks

The NBL is strongly affected by inhomogeneities in the underlying surface. Under the conditions encountered in central Oklahoma (i.e., rolling hills of 20 to 30 m height on a horizontal scale of 1 to 10 km), we have found from aircraft observations and demonstrated theoretically that the flow contains stationary waves that are fixed with respect to the terrain. Aircraft are, in some respects, well-suited to probing spatially-varying boundary-layer structure, although the shallowness and vertical variability of the NBL limits the ability of an aircraft to measure detailed structure near the surface. Perhaps even more than in the convective boundary layer, a combined aircraft-tower experiment, similar to that reported by Kaimal *et al.* (1982) for the convective boundary layer, would be useful in investigating both the temporal and spatial variability of the NBL.

Acknowledgements

We thank Richard Rotunno for discussing with us the types of waves that may exist in the NBL, and the personnel at the NCAR Research Aviation Facility for their assistance during and after the experiment. We also thank C. Nappo, J. Wyngaard, L. Mahrt, Z. Sorbjan, and M. Raupach for their helpful comments.

References

André, J. C. and Mahrt, L.: 1982, 'The Nocturnal Surface Inversion and Influence of Clear-Air Radiative Cooling', *J. Atmos. Sci.* **39**, 864–878.

Caughey, S. J., Wyngaard, J. C., and Kaimal, J. C.: 1979, 'Turbulence in the Evolving Stable Layer', *J. Atmos. Sci.* **36**, 1041–1052.

Clarke, R. H., Dyer, A. J., Brooks, R. R., Reid, D. G., and Troup, A. J.: 1971, 'The Wangara Experiment: Boundary-Layer Data', *Technical Paper No. 19*, Division of Meteorological Physics, CSIRO, Aspendale, Australia, 340 pp.

Gossard, E. E. and Hooke, W. H.: 1975, *Waves in the Atmosphere*, Elsevier Scientific Publ. Co., New York, p. 456.

Hunt, J. C. R., Kaimal, J. C., and Gaynor, J. E.: 1985, 'Some Observations of Turbulence Structure in Stable Layers', *Quart. J. R. Meteorol. Soc.* **111**, 793–815.

Kaimal, J. C., Eversole, R. A., Lenschow, D. H., Stankov, B. B., Kahn, P. H., and Businger, J. A.: 1982, 'Spectral Characteristics of the Convective Boundary Layer Over Uneven Terrain', *J. Atmos. Sci.* **39**, 1098–1114.

Mahrt, L.: 1985, 'Vertical Structure and Turbulence in the Very Stable Boundary Layer', *J. Atmos. Sci.* **42**, 2333–2349.

Rotunno, R.: 1977, 'Internal Gravity Waves in a Time-Varying Stratification', *J. Fluid Mech.* **82**, part 4, pp. 609–619.

Acknowledgements

[illegible]

References

[illegible]

SPECTRA OF VELOCITY CIRCULATION IN THE SURFACE LAYER OF THE ATMOSPHERE

B. M. KOPROV, G. V. AZIZYAN, and V. V. KALUGIN

Institute of Atmospheric Physics, USSR Academy of Sciences, Moscow, U.S.S.R.

(Received 29 August, 1987)

Abstract. Measurements of the velocity circulation in the atmospheric surface layer can be used to infer information on atmospheric vorticity, a quantity of fundamental importance. In this paper, circulation is measured with an acoustic sensor. The instrument is described, and some results are given.

1. Introduction

Measurements of velocity circulation have been made to obtain information on the vorticity, $\mathbf{\Omega} = \mathrm{curl}\,\mathbf{v}$ (Bovsheverov *et al.*, 1971; Ohtou *et al.*, 1983; Elefterios *et al.*, 1983). The method of measurement was based on Stokes's theorem, relating vorticity flux to velocity circulation. There are two major reasons for interest in vorticity near the surface. First, vorticity is the most specific feature of the turbulent flow. Second, for isotropic turbulence there is a relationship between absolute vorticity ($|\mathbf{\Omega}|$) and the rate of dissipation of turbulent energy: $\varepsilon = \nu |\mathbf{\Omega}|^2$, where ν is kinematic viscosity (Landau and Lifshitz, 1986). The mean value of ε is known to be the most important characteristic of turbulent flow and statistics of ε-variability must be taken into account for a precise description of a turbulent velocity field (Yaglom, 1966).

The correlation radius of the vorticity field in isotropic turbulence is of the same order as the Kolmogorov microscale $\eta = (\nu^3/\varepsilon)^{1/4}$, which is typically less than 1 mm. Therefore, to quantify any vorticity component by measuring the velocity circulation, the contour size must be less than η. Thermoanemometers are sufficiently small that it is possible to measure the streamwise vorticity component with 1 mm resolution (Elefterios *et al.*, 1983; Wyngaard, 1969).

It has recently been reported that turbulent flow near a boundary produces so-called coherent structures, i.e., 'connected large-scale turbulent fluid mass with a phase-correlated vorticity over its spatial extent' (Hussain, 1983). To investigate such structures experimentally, there is no need to use a very small sensor, as would be necessary to resolve fluctuations of size η. An alternative approach is to measure the velocity circulation with an acoustic sensor. In this case, every component of the vorticity vector may be measured, but it is difficult to make the contour size less than 1 cm. In the present paper, we describe an acoustic sensor developed to measure velocity circulations near the surface, and we present data on circulation spectra for different atmospheric stabilities and heights from 1 to 5 m above the surface.

Boundary-Layer Meteorology **42** (1988) 137–143.

The spectra differ from those calculated for an isotropic velocity field. Possible methods for deriving information on vorticity and vorticity flux in the atmospheric surface layer are discussed.

2. Instrumentation and Field Measurements

The operating principles of the velocity circulimeter sensor have been presented previously (Bovsheverov *et al.*, 1971). The main components of the sensor (Figure 1) are

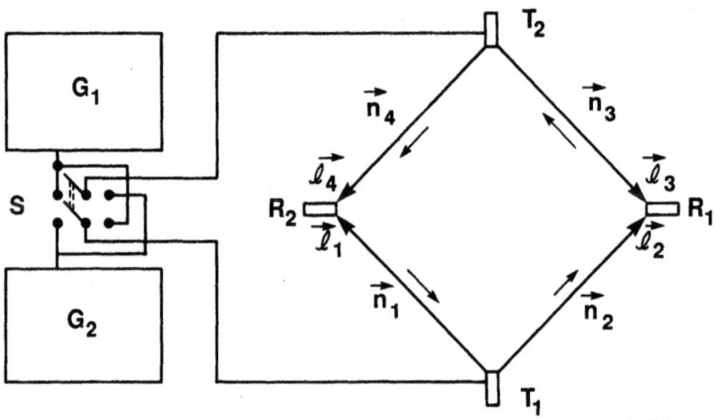

Fig. 1. A sketch of the velocity circulation sensor. G_1 and G_2 are generators of sinusoidal signals with frequencies f_1 and f_2, T_1, and T_2 are transmitters, R_1 and R_2 are receivers, and S is a switch. Vectors $\mathbf{l}_1 - \mathbf{l}_4$ indicate the acoustic beams, vectors $\mathbf{n}_1 - \mathbf{n}_4$ denote directions of moving around the contour.

four acoustic transducers, two transmitters, and two receivers. Sinusoidal signals from the generators G_1 and G_2 are transmitted with frequencies f_1 and f_2 and then received by each receiver. After amplifying and detecting the signals from the two receivers, we obtain a sinusoidal signal at beat frequency $(f_1 - f_2)$. The phase of the sinusoid at the beat frequency contains the desired information. As shown by Bovsheverov *et al.* (1971), the phase difference between these two signals can be written as follows:

$$\Delta\phi = \frac{\omega_1}{C_0^2} (\Gamma + \lambda + V),$$

where $\omega = 2\pi f$; C_0 is mean velocity of sound; C_i is the velocity of sound averaged over the ith side of the sensor square contour; Γ is the velocity circulation; $\lambda = (C_2 - C_1) + (C_4 - C_3)$; and $V = (\Delta\omega/\omega)\,[(C_2 - C_1) + v_1 + v_4]$, where v_1 and v_4 are the velocity projections on the vectors $\mathbf{n}_1$ and $\mathbf{n}_4$.

Bruel and Kyer condenser microphones were used both for transmitting and receiving the acoustic signals. Frequencies were chosen to be 80 and 75 kHz. The noise level of phasemeter was of order of 0.2°.

It is important to consider possible errors caused by distortion of the flow by the sensor. The best aerodynamic conditions apply in the case of measurement of Γ_1, because the plane of the contour in this case is perpendicular to the wind direction and the aerodynamic shadow lies outside the measuring contour. The best situation for measuring circulations Γ_2 and Γ_3, with contours in the plane of mean velocity vector, is shown in Figure 1. To minimize the magnitude in the V signal, if measuring the circulation Γ_2, it is desirable to arrange the sensor such that the line connecting the two receivers is horizontal.

The measurements reported here were made in Tsimlyansk in August 1985. The weather was clear and dry, and the wind was steady. Information on the measuring conditions and some results are presented in Table I. Values of the friction velocity u_* and temperature scale

$$T_* = -\frac{\overline{w'T'}}{ku_*}$$

(where w' and T' are fluctuations of the vertical wind component and temperature, respectively, and $k = 0.4$ is the von Kármán constant) were calculated from the wind and temperature profiles. The fine-scale sensors were mounted on a pantograph mast. Data were processed and analyzed using a special assemblage of computing equipment, based on the Soviet-made Eletronika-60 computer.

TABLE I

un umber	Date	Time	Altitude (m)	Sensor orientation	u (m s^{-1})	T (°C)	V (m s^{-1})	T (°C)	ζ (10^2)	P
3	26.08	13–06	1	2	3.3	24.2	0.32	−0.29	−1.48	+0.28
5	27.08	11–45	1	2	4.9	24.6	0.45	−0.31	−0.79	+0.27
6	27.08	12–11	1	2	4.9	25.0	0.47	−0.31	−0.73	+0.27
7	27.08	12–57	2	2	5.6	25.7	0.49	−0.31	−1.36	+0.34
3	27.08	13–18	5	2	6.4	25.4	0.50	−0.31	−3.27	+0.66
[illegible]	27.08	13–51	1.6	2	5.0	26.8	0.40	−0.31	−1.62	+0.30
[illegible]	27.08	14–22	0.6	2	4.7	27.0	0.43	−0.29	−0.47	+0.24
[illegible]	28.08	11–47	0.9	3	4.9	27.2	0.53	−0.35	−0.59	+0.31
2	28.08	12–05	0.9	3	4.7	27.7	0.50	−0.29	−0.54	+0.31
3	28.08	12–27	1.85	3	5.5	28.1	0.56	−0.29	−0.94	+0.46
4	28.08	12–50	5.1	3	6.9	27.2	0.51	−0.28	−2.87	+0.84
[illegible]	29.08	18–47	1.85	2	4.2	29.9	0.43	−0.02	−0.11	+0.41
2	29.08	19–09	1.85	2	3.9	29.3	0.37	−0.00	−0.00	+0.23
[illegible]	30.08	11–11	1.35	3	5.4	26.0	0.54	−0.26	−0.66	+0.54
[illegible]	30.08	12–16	1.8	1	5.6	28.6	0.51	−0.26	−0.94	+0.37
7	30.08	19–17	1.8	1	3.9	27.2	0.34	−0.00	−0.00	+0.31
[illegible]	30.08	20–09	1.8	1	4.4	26.1	0.52	+0.04	+0.14	+0.43
[illegible]	31.08	11–19	5.2	1	6.1	24.6	0.40	−0.22	−3.74	+0.72
[illegible]	31.08	11–47	5.2	1	5.0	25.4	0.34	−0.25	−5.76	+0.62
[illegible]	31.08	12–14	5.2	1	5.3	26.2	0.36	−0.22	−4.6	+0.62

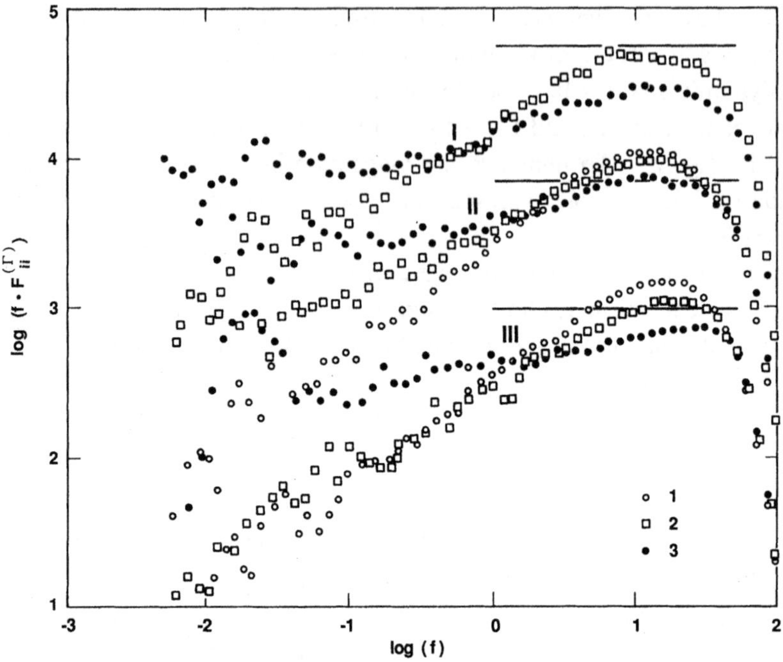

Fig. 2. Spectra of velocity circulation for different orientations of the sensing contour in unstable conditions. (I) Γ_1, runs 26, 29; (II) Γ_2, runs 5, 7, 8; (III) Γ_3, runs 12, 13, 14; (1) $z = 1$ m; (2) $z = 2$ m; (3) $z = 5$ m. The horizontal lines correspond to equal levels of spectral densities.

3. Results

Figure 2 presents the product of f and the spectral density of the velocity circulation at the ith orientation ($F_{ii}^{(\Gamma)}(f)$). These data were obtained in unstable conditions at heights of 1, 2, and 5 m. The index $i = 1$ corresponds to the case for which the vector normal to the contour plane is parallel to the wind direction, $i = 2$ indicates that this vector is horizontal and perpendicular to the wind, $i = 3$ refers to the case in which the normal vector is directed vertically.

Inspection of Figure 2 reveals that there is an interval from 0.1 to about 4 Hz in which all functions approximately obey a power law with an exponent $\alpha < 1$. At higher frequencies, deviation from this behavior is evident. This deviation is caused by spatial averaging and starts from the point $k_1 l \sim 1$, where $k_1 = 2\pi f/\bar{u}$, l is the contour size, and $\bar{u}$ is the mean wind velocity.

Figure 2 shows that spectral slopes are different at heights of 1, 2, and 5 m. The stability dependence of the values p_1, p_2, and p_3 ($p_i = 1 - \alpha_i$) is shown in Figure 3; here, the values are plotted against $\zeta = z/L$, where L is the Monin–Obukhov length scale. These data indicate that circulation spectra do not obey a universal law in the investigated range: exponents p_i depend on both ζ and i. The stability index ζ is the ratio of thermal to dynamic contributions to kinetic energy generation. Therefore, the growth

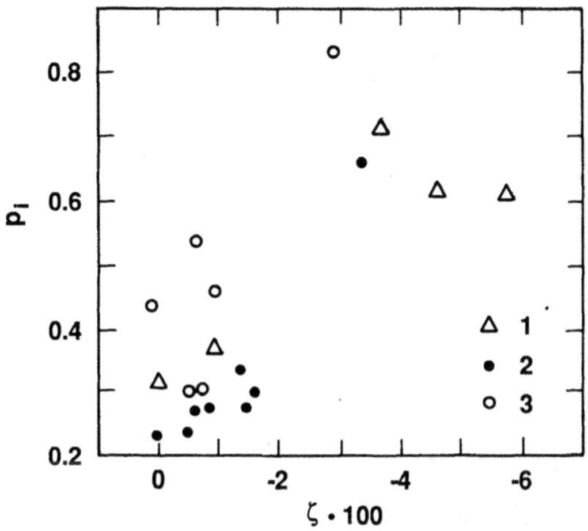

Fig. 3. The dependence of the exponents ($p_i = 1 - \alpha_i$) on stability for different orientations of the sensor. $i = 1$: longitudinal; $i = 2$: lateral; $i = 3$: vertical.

of p_i with increasing $|\zeta|$ in unstable conditions may be interpreted as an indication of the fundamental role of the convective mechanism on the generation of large-scale vortices.

The exponent p_3 is much greater than p_2 and p_1; this is probably due to a decreased influence of the earth's surface on big vortices with vertical axis larger than on the others. If this is really so, then vortices with a size much greater than Z must be almost two-dimensional.

The present data make it possible to estimate not only the slopes of spectral functions, but also the ratio between the spectral densities at different heights. It turns out that the intensity of vorticity fluctuations in the vicinity of 2 Hz is almost equal at all heights; at lower frequencies, the spectral density increases with height, and at higher ones, it decreases. It is to be noted that estimates of the exponent p are not greatly affected by errors λ and V. To minimize any such errors, we made use of the following features of the circulimeter: (1) switching the generators by switch S (Figure 1) causes Γ and V to change sign, but λ does not, (2) when rotating the sensor around the axis perpendicular to its plane, Γ and λ keep their sign, but V changes sign.

The findings were compared with those computed for locally isotropic turbulence. One-dimensional spectra of projections of vorticity on the plane square contour with the size 'l' oriented along and perpendicular to the x_1 direction were calculated along the direction x_1. For this purpose, three-dimensional velocity spectra were presumed to be described by

$$E(k) = c_1 \varepsilon^{2/3} k^{-5/3} \phi_\varepsilon(k\eta),$$

where

$$k = (k_1^2 + k_2^2 + k_3^2)^{1/2}$$

and

$$\phi_\varepsilon(k\eta) = \begin{cases} 1 & \text{for} \quad k\eta \leq 1\,, \\ 0 & \text{for} \quad k\eta > 1\,. \end{cases}$$

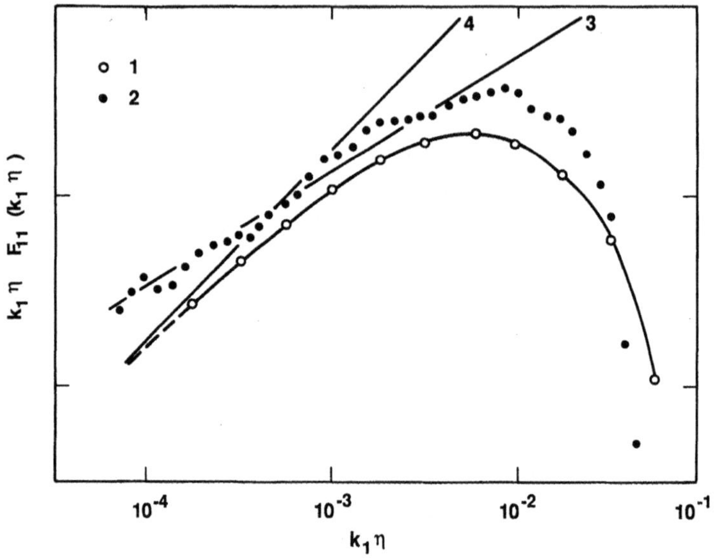

Fig. 4. Comparison of measured velocity circulation spectra (run 27) with predictions for isotropic turbulence. (1) computed, $l/\eta = 200$, (2) measured, $l/\eta = 170$. Straight lines 4 and 3 correspond to $p_1 = 0$ and $p_1 = 0.41$, respectively.

Spectra $F_{ii}^{(\Gamma)}(k, l, \eta)$ for lateral and longitudinal orientation of the contour and for 15 values of parameter l/η, between the values $\frac{1}{40}$ and 400, were computed numerically.

Figure 4 allows us to compare the behaviour of the theoretical function $k_1 F_{11}^{(\Gamma)}(k, \eta)$ for $l/\eta = 200$ with that measured at a height of 2 m in neutral conditions with $l/\eta = 170$. Both functions increase with increasing $k_1\eta$ up to $k_1\eta = 0.007$, and then decrease. The straight line drawn in the diagram corresponds to the case $F_{11}^{(\Gamma)}(k, \eta) = \text{const}$. More rapid drop-off of the experimental function for $k_1\eta > 0.007$ was caused by the phasemeter filter properties.

As one can see, the experimental and theoretical functions behave noticeably differently, in contrast to the conclusion of Bovsheverov *et al.* (1971). It is possible that this apparent disagreement is a consequence of the present use of a different method of data presentation: we used a log-log plot, whereas Bovsheverov *et al.* used log-linear. The calculated spectra are illustrated in Figure 5.

4. Conclusions

The data presented in this paper indicate that spectra of circulation differ noticeably from those predicted by isotropic theory; in particular, the relative contribution of the

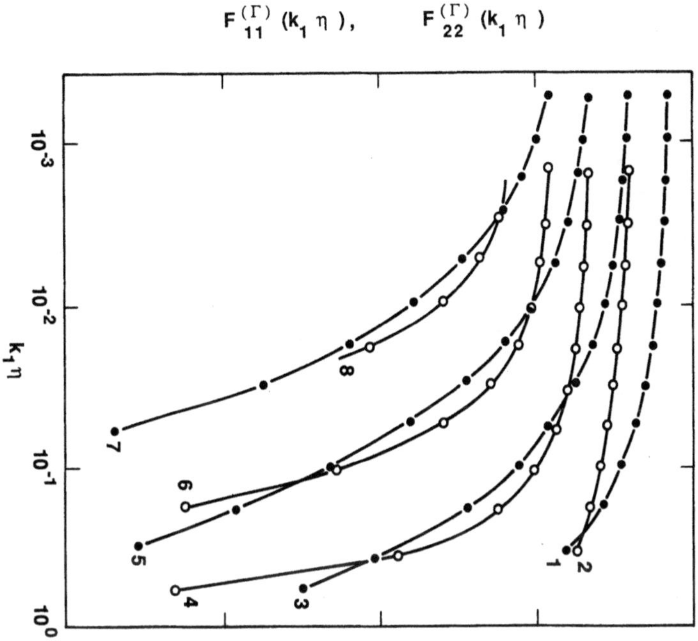

Fig. 5. Dependence of one-dimensional spectra of velocity circulation on the parameter l/η for the longitudinal and lateral orientation of the contour in isotropic turbulence. Even numbers – longitudinal, odd ones – lateral. 1, 2: $l/\eta = 0.1$; 3, 4: $l/\eta = 12.5$; 5, 6: $l/\eta = 50$; 7, 8: $l/\eta = 200$.

low-frequency components to the measured circulation spectra is much more than that for the isotropic case.

This means that we can consider the measured circulation spectra as an approximation to the vorticity spectra. In addition, it seems reasonable to think that the flux of vorticity in the atmospheric surface layer is caused mainly by large-scale coherent structures. Hence, one-point combination of the velocity circulation sensor with those of the velocity component and temperature allows the low-frequency part of corresponding cospectra (Ohtou *et al.*, 1983) to be obtained.

References

Bovsheverov, V. M., Gurvich, A. S., Kochetkov, A. N., and Lomadze, S. O.: 1971, 'Measurement of Frequency Spectrum of Small-Scale Velocity Circulation in a Turbulent Flow', *Izv. USSR Acad. of Sciences, Phys. Atm. and Ocean* **7**, 371–376.

Elefterios, G., Kastrinakis, and Eckelmann, H.: 1983, 'Measurement of Streamwise Vorticity Fluctuations in a Turbulent Channel Flow', *J. Fluid Mech.* **137**, 165–186.

Hussain, A. K. M. F.: 1983, 'Coherent Structures – Reality and Myth', *Phys. Fluids* **26**, 2816–2838.

Landau, L. D. and Lifshitz, E. M.: 1986, *Hydrodynamics M.*, Nauka, Moscow, 756 pp.

Ohtou, A., Maitani, T., and Seo, T.: 1983, 'Direct Measurement of Velocity and Its Transport in the Surface Layer Over a Paddy Field', *Boundary-Layer Meteorol.* **27**, 197–207.

Wyngaard, J. C.: 1969, 'Spatial Resolution of the Vorticity Meter and Other Hot-Wire Arrays', *J. Sci. Instr.* **9**, 983–987.

Yaglom, A. M.: 1966, 'On the Influence of the Dissipation Energy Fluctuations on the Characteristics of Turbulence in the Inertia Interval', *Doklady of USSR Acad. of Sciences* **186**, 49–52.

A NOTE ON THE BUSINGER–DYER PROFILES

(Research Note)

J. A. BUSINGER

*National Center for Atmospheric Research, Boulder, Co., U.S.A.**

(Received 23 July, 1987)

Symbols

K_h eddy thermal diffusivity,
K_m eddy viscosity,

L Obukhov length $\equiv -\dfrac{u_*^3 \overline{T}}{kg\overline{w' T_v'}}$,

where u_* is friction velocity,
T is absolute temperature,
T_v is virtual temperature,
k is von Kármán constant,
g is acceleration due to gravity,

Ri Richard number $\equiv g/\theta \dfrac{\partial\overline{\theta}/\partial z}{(\partial\overline{u}/\partial z)^2}$,

where u is the horizontal velocity component,
z the height above the surface,
θ is potential temperature,

ϕ_m dimensionless wind gradient $\equiv \dfrac{kz}{u_*}\dfrac{\partial\overline{u}}{\partial z}$,

ϕ_h dimensionless temperature gradient $\equiv \dfrac{kz}{u_*}\dfrac{\partial\overline{\theta}}{\partial z}$,

where $\theta_* \equiv -\dfrac{\overline{w'T'}}{u_*}$,

γ a constant ~ 16,
ζ dimensionless height $\equiv z/L$.

* The National Center for Atmospheric Research is sponsored by the National Science Foundation.

1. Introduction

In the micrometeorological literature, reference is sometimes made to the 'Businger–Dyer Profiles' or the 'Dyer–Businger profiles/relations' without referring to the origin of these relations. For example, in the textbook on 'Atmospheric Turbulence' by Panofsky and Dutton (1984) on p. 134, reference is made to the 'Businger–Dyer formula'. To add to the mystery, these authors refer on p. 141 to the Businger–Dyer–Pandolfo empirical result (Businger, 1966; Pandolfo, 1966) that in unstable air

$$\mathrm{Ri} = \frac{z}{L} \equiv \zeta\,. \tag{1}$$

So it seemed to me that it would be appropriate for this issue of *Boundary-Layer Meteorology* which is dedicated to Arch Dyer, to go back to 1965 and describe the circumstances that led to the above mentioned profiles as I remember them.

2. Aspendale, 1965

In the academic year 1965–1966 I found myself in Australia on a sabbatical leave. The first part of this leave was spent at the CSIRO Division of Meteorological Physics in Aspendale, Victoria. It was in many ways a good choice. The fall is exchanged for spring, SE Australia is a pleasant place to be in spring and summer, and the scientists in the division were hospitable and stimulating. Priestley, Swinbank, Dyer, Webb, McIlroy, Taylor, Clarke, Deacon and several others carried out an active research program.

I intended to work on one specific problem in turbulence, with the hope of solving it. After a few days in the library, I found out that the problem in question had already been solved. Moreover, the solution was so elegant that it made me feel quite humble. On the other hand, this left me completely open on what to do next. This gave me a wonderful sense of freedom to explore whatever struck my fancy. Thus I started to look at the excellent surface-layer data that had been collected in recent years by Swinbank and Dyer near Kerang and Hay.

One sunny afternoon, Bill Swinbank came into my office and showed me the precursor to Figure 1, a good confirmation of the Monin–Obukhov similarity. I agreed that it was an exciting graph on which to meditate. The plot of the Richardson number, Ri versus $\zeta \equiv z/L$ (where L is the Obukhov length) is a basic test of the Monin–Obukhov similarity because Ri contains only information about gradients and ζ contains only information about fluxes. Looking at the graph, it struck me that the relation was close to a one-to-one relation, i.e.,

$$\mathrm{Ri} = \zeta\,. \tag{1}$$

This impression was strengthened by the fact that I believed that ζ was overestimated because u_* was underestimated. Dyer (1967) also comments on the underestimation

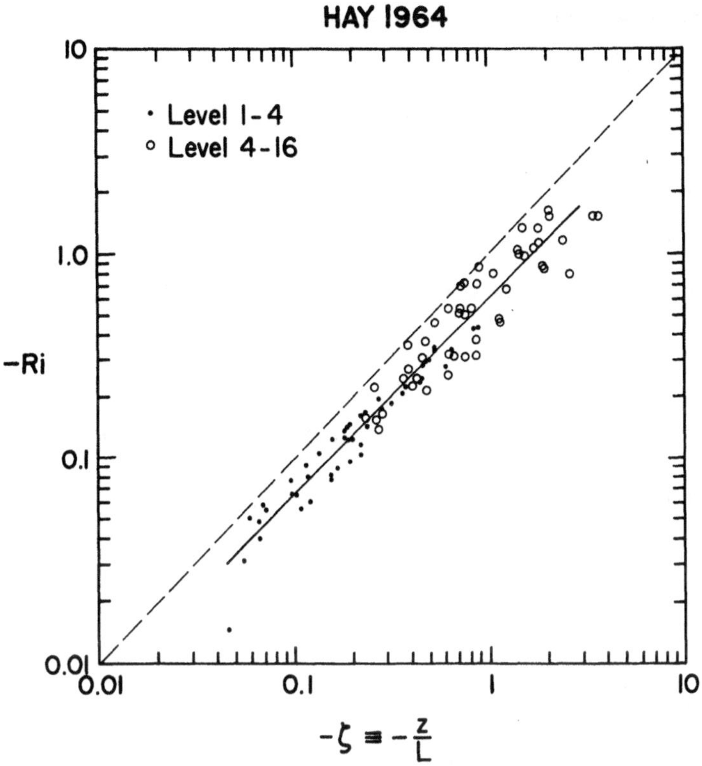

Fig. 1. Ri versus − ζ. Hay observations (after Swinbank, 1968).

of u_*. The reason for this notion stemmed from the fact that the stress had not been measured directly but was derived by applying a drag coefficient to the lowest windspeed measurement in the profile. The drag coefficient was obtained during neutral conditions when the logarithmic profile is valid. Under unstable conditions, the profile is no longer logarithmic but increases less rapidly with height. So if the lowest observed windspeed is an underestimate of the neutral windspeed with the same stress, the neutral drag coefficient applied to the observed wind yields a stress value that is too low and consequently u_* is too low. A 10% error in u_* translates into a 30% error in ζ. Such a correction would bring Swinbank's regression line very close to the one-to-one relation.

Whether or not this reasoning was valid is immaterial at this point. What matters is that it started a speculation on what the consequences would be of Equation (1) for the profile descriptions.

From the definitions of Ri, ζ, ϕ_m, and ϕ_h, we have the identity

$$\mathrm{Ri} = \frac{\phi_h}{\phi_m^2}\,\zeta\,, \qquad (2)$$

which combined with (1) leads to

$$\phi_h = \phi_m^2 \tag{3}$$

and furthermore, because $\phi_h/\phi_m = K_h/K_m$ we find that

$$\frac{K_h}{K_m} = \phi_m^{-1} . \tag{4}$$

Using a simple mixing-length model (Fleagle and Businger, 1963 or 1980), I had derived the relation

$$\phi_m = (1 - \gamma\,\mathrm{Ri})^{-1/4} . \qquad (5)^*$$

This combined with (1), (3), and (4) leads to the Businger–Dyer profiles:

$$\phi_m = (1 - \gamma\zeta)^{-1/4} , \tag{6}$$

$$\phi_h - (1 - \gamma\zeta)^{-1/2} , \tag{7}$$

and

$$\frac{K_h}{K_m} = (1 - \gamma\zeta)^{1/4} . \tag{8}$$

The result of (6)–(8) was quite exciting to me, so I showed it around. First I showed it to Swinbank. He was not very interested in this new profile formulation because he had just published the 'exponential profile' (Swinbank, 1964) based on an elegant derivation; however, the fit to the data was poor. Then I went to Arch Dyer and much to my surprise he had written the same Equations (6)–(8) on the blackboard in his office. Apparently he had been intrigued by the same plot of Ri versus ζ and came to the same conclusion. In fact, in a 1964 paper, he had analyzed heat flux data and found the same power law as (7) indicates. So, it is possible that his reasoning went from (1), (2) and (7) to (6) and (8). I regret not having asked him this question at the time.

Arch and I discussed briefly the idea of putting these profiles in a joint paper and publishing them. Unfortunately, Bill Swinbank was unhappy about competing profiles and felt that the staff of the Division should show some 'loyalty' and not prematurely publish the above results. Consequently Arch refrained from working on a joint paper

* Equation (5) combined with (2) leads to the KEYPS or O'KEYPS function (Panofsky, 1963; Businger and Yaglom, 1971), i.e.,

$$\phi_m^4 - \gamma'\zeta\phi_m^3 = 1 ,$$

where

$$\gamma' = \gamma \frac{K_h}{K_m} .$$

This function therefore requires thatK_h/K_m = constant. O'KEYPS stands for the initials of the various authors who derived this equation independently.

and I published the profiles as part of the proceedings of a meeting organized by the Rand Corporation (Businger, 1966). In a completely independent study, Pandolfo (1966) also arrived at (1).

3. How Good is the Assumption Ri = ζ?

It is of interest to see how well the assumption (1) has been verified by later experiments. Dyer and Hicks (1970) reported results of experiments at Hay (New South Wales) described by Swinbank and Dyer (1967) and at Gurley (New South Wales). In these experiments, the momentum flux was measured directly with the 'fluxatron' (Dyer *et al.*, 1967; Hicks, 1970). The results presented in Figure 2 suggest that (1) fits the data very well, indeed.

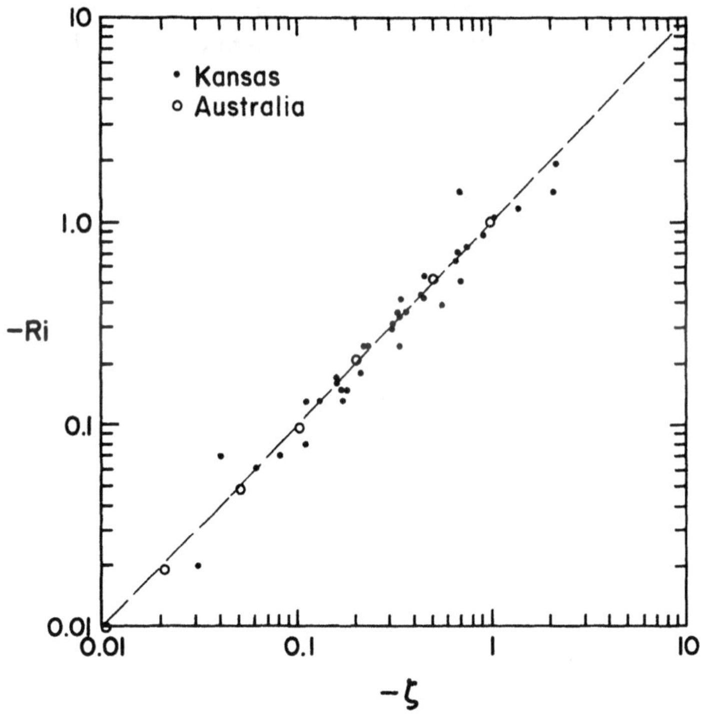

Fig. 2. As Figure 1 for Kansas observations. ● (Izumi, 1971), and Australian observations, ○ (Dyer and Hicks, 1970).

In the same period, the Boundary Layer Group of AFCRL (Air Force Cambridge Research Laboratories) carried out an experiment in Kansas. Again fluxes and profiles were measured independently. Results concerning (1) have been reported by Businger *et al.* (1971) and Izumi (1971). These results are also presented in Figure 2, again suggesting that (1) is a good assumption. The scatter is greater in this set because individual runs are shown, rather than the average of a series of runs.

The two data sets differ in several ways. The fluxatron used by Dyer and Hicks has a lower response time than the sonic anemometer used in Kansas and therefore a correction to the co-spectrum is needed. In the Kansas experiment no vapor flux was measured; therefore ζ was slightly underestimated. The von Kármán constant, k, that fit the Australian data best was 0.41, whereas the Kansas data suggested $k = 0.35$. Thus, in some sense the agreement between the two data sets in Figure 2 must be considered fortuitous.

More recent efforts to look at the relation $\mathrm{Ri} = f(\zeta)$ show a continuation of the trend between Figures (1) and (2). Dyer and Bradley (1982) give an analysis of the ITCE (International Turbulence Comparison Experiment) data taken in 1976. The result is shown in Figure 3. We see that in this case the relation is to the left of the one-to-one line, almost as much as Swinbank's (Figure 1) is to the right of this line. Webb (1982) reanalyzed the old data taken before 1965. Surprisingly this analysis gives almost the same $\mathrm{Ri} = f(\zeta)$ function as Dyer and Bradley find; see Figure 3. From these figures, we see that over the last 20 years $-$ Ri has steadily increased versus $-\zeta$. Is this a continuing trend, or is it part of an oscillation? As it stands, the differences between the various data sets and the analyses are still too large to feel confident of the results. Mortensen (personal communication) recently showed me preliminary results from a large data set taken in Denmark. These preliminary results agree with the analysis of Dyer and Bradley (1982).

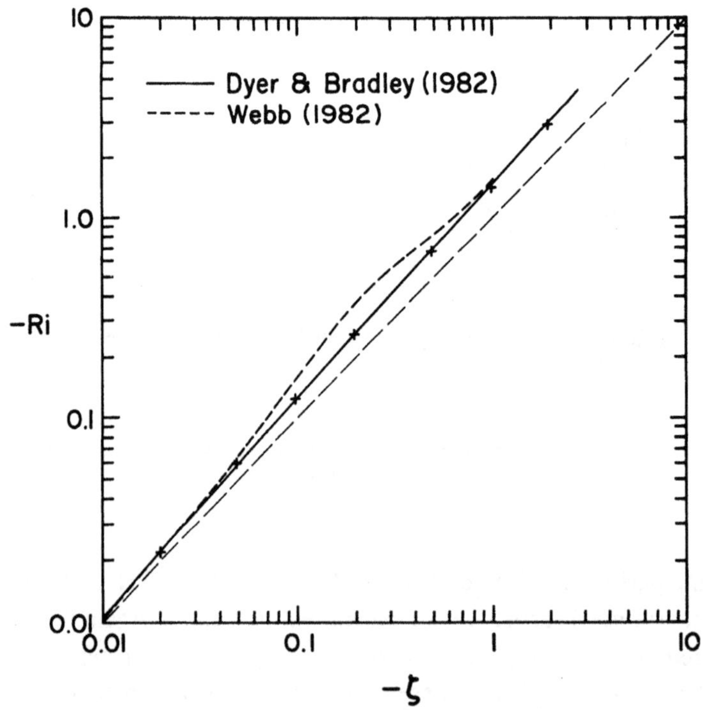

Fig. 3. As Figure 1 + according to Dyer and Bradley (1982) and dashed line according to Webb (1982).

To settle this and other issues on the structure of the atmospheric surface layer, more and better experimental data are needed. To paraphrase Dyer and Bradley (1982): "The enormous data scatter which is typical of this type of experiment points to the need for large amounts of data in order to obtain a statistically significant result... The evidence is mounting that the atmosphere does not follow the averaged laws at all places and at all times, even over an excellent site..." Mesoscale structures imbedded in the planetary boundary layer may, to a large extent, contribute to this large scatter. Therefore, the next generation of surface-layer experiments should be conducted in a setting where adequate documentation of the entire boundary layer is available.

References

Businger, J. A.: 1966, 'Transfer of Momentum and Heat in the Planetary Boundary Layer', *Proceedings of the Symposium on Arctic Heat Budget and Atmospheric Circulation*, The Rand Corporation, pp. 305–332.

Businger, J. A. and Yaglom, A. M., 1971, 'Introduction to Obukhov's Paper on 'Turbulence in an Atmosphere with a Non-Uniform Temperature'', *Boundary-Layer Meteorol.* **2**, 3–6.

Businger, J. A., Wyngaard, J. C., Izumi, Y., and Bradley, E. F.: 1971, 'Flux-Profile Relationships in the Atmospheric Surface Layer', *J. Atmos. Sci.* **28**, 181–189.

Dyer, A. J.: 1965, 'The Flux-Gradient Relation for Turbulent Heat Transfer in the Lower Atmosphere, *Quart. J. Roy. Meteorol. Soc.* **91**, 151–157.

Dyer, A. J.: 1967, 'The Turbulent Transport of Heat and Water Vapour in an Unstable Atmosphere', *Quart. J. Roy. Meteorol. Soc.* **93**, 501–508.

Dyer, A. J. and Bradley, E. F.: 1982, 'An Alternative Analysis of Flux-Gradient Relationships at the 1976 ITCE', *Boundary Layer Meteorol.* **22**, 3–19.

Dyer, A. J., Hicks, B. B., and King, K. M.: 1967, 'The Fluxatron – A Prevised Approach to the Measurement of Eddy Fluxes in the Lower Atmosphere', *J. Appl. Meteorol.* **6**, 408–413.

Dyer, A. J. and Hicks, B. B.: 1970, 'Flux-Gradient Relationships in the Constant Flux Layer', *Quart. J. Roy. Meteorol. Soc.* **96**, 715–721.

Fleagle, R. G. and Businger, J. A.: 1963 (2nd ed., 1980): *An Introduction to Atmospheric Physics*, Academic Press, New York, 356 pp. (432 pp).

Hicks, B. B.: 1970, 'The Measurement of Atmospheric Fluxes Near the Surface: A Generalized Approach', *J. Appl. Meteorol.* **9**, 386–388.

Izumi, Y.: 1971, 'Kansas 1968 Field Program Data Report', *Environmental Research Papers*, No. 379 Meteor. Lab. AFCRL, Air Force Systems Command, USAF.

Pandolfo, J. P.: 1966, 'Wind and Temperature Profiles for Constant-Flux Boundary Layers in Lapse Conditions with a Variable Eddy Conductivity to Eddy Viscosity Ratio', *J. Atmos. Sci.* **23**, 495–502.

Panofsky, H. A.: 1963, 'Determination of Stress from Wind and Temperature Measurements', *Quart. J. Roy. Meteorol. Soc.* **89**, 85–94.

Panofsky, H. A. and Dutton, J. A.: 1984, *Atmospheric Turbulence*, J. Wiley and Sons, New York, 397 pp.

Swinbank, W. C.: 1964, 'The Exponential Wind Profile', *Quart. J. Roy. Meteorol. Soc.* **90**, 119–135.

Swinbank, W. C.: 1968, 'A Comparison Between Predictions of Dimensional Analysis for the Constant-Flux Layer and Observations in Unstable Conditions', *Quart. J. Roy. Meteorol. Soc.* **94**, 460–467.

Swinbank, W. J. and Dyer, A. J.: 1967: 'An Experimental Study in Micro-Meteorology', *Quart. J. Roy. Meteorol. Soc.* **93**, 494–500.

Webb, E. K.: 1982, 'Profile Relationships in the Super Adiabatic Surface Layer', *Quart. J. Roy. Meteorol. Soc.* **108**, 661–688.

ANNOUNCEMENT

TOPICS IN MICROMETEOROLOGY

A Festschrift for Arch Dyer

Guest Editor: BRUCE B. HICKS

Please note that a hardbound edition of these special issues of *Boundary-Layer Meteorology*, Vol. 42, Nos. 1–2 (January 1988) is available from the publishers.

ISBN 90–277–2694–9 Price: Dfl. 125,– $69.00 £38.00

GPSR Compliance

The European Union's (EU) General Product Safety Regulation (GPSR) is a set of rules that requires consumer products to be safe and our obligations to ensure this.

If you have any concerns about our products, you can contact us on ProductSafety@springernature.com

In case Publisher is established outside the EU, the EU authorized representative is:

Springer Nature Customer Service Center GmbH
Europaplatz 3
69115 Heidelberg, Germany

Batch number: 10584448

Printed by Printforce, the Netherlands